New Approaches on Energy and the Environment

New Approaches on Energy and the Environment: Policy Advice for the President

Edited by
Richard D. Morgenstern
and
Paul R. Portney

Resources for the Future | Washington, DC

An RFF Press book
Published by Resources for the Future
1616 P Street NW
Washington, DC 20036–1400
USA
www.rffpress.org

Library of Congress Cataloging-in-Publication Data
New approaches on energy and the environment : policy advice for the president / edited by: Richard D. Morgenstern and Paul R. Portney.
 p. cm.
 ISBN 1-933115-00-9 (cloth : alk. paper) — ISBN 1-933115-01-7 (pbk. : alk. paper)
 1. Energy policy—United States. 2. Environmental policy—United States. 3. Environmental protection—United States. I. Morgenstern, Richard D. II. Portney, Paul R.
 HD9502.U52N45 2004
 333.79'0973'09051—dc22   2004018822

f e d c b a

The paper in this book meets the guidelines for permanence and durability of the Committee on Production Guidelines for Book Longevity of the Council on Library Resources. This book was typeset in Utopia and Ocean Sans by Peter Lindeman. It was copyedited by Joyce Bond. The cover was designed by Marc Alain Meadows, Meadows Design Office, Inc. Cover art: Dave Cutler Studio.

ISBN 1-933115-00-9 (cloth)   ISBN 1-933115-01-7 (paper)

About *Resources for the Future* and *RFF Press*

Resources for the Future (RFF) improves environmental and natural resource policy-making worldwide through independent social science research of the highest caliber. Founded in 1952, RFF pioneered the application of economics as a tool for developing more effective policy about the use and conservation of natural resources. Its scholars continue to employ social science methods to analyze critical issues concerning pollution control, energy policy, land and water use, hazardous waste, climate change, biodiversity, and the environmental challenges of developing countries.

RFF Press supports the mission of RFF by publishing book-length works that present a broad range of approaches to the study of natural resources and the environment. Its authors and editors include RFF staff, researchers from the larger academic and policy communities, and journalists. Audiences for publications by RFF Press include all of the participants in the policymaking process—scholars, the media, advocacy groups, NGOs, professionals in business and government, and the public.

Contents

Part III. Natural Resources

Part IV. Information Decision Frameworks

Foreword

The policy recommendations in this collection were written in advance of the outcome of the 2004 presidential election, at a time when energy and the environment were campaign issues overmatched by economic and international developments. Although the two party platforms clearly differed in their approaches to energy and environmental policy, this book addresses a range of meritorious ideas that both presidential candidates might agree are worth serious consideration. The fact that we did not know who would be victorious—and thus could not skew our recommendations toward one candidate or the other—should give the ideas presented here even more credibility.

Moreover, the fact that the two of us jointly and enthusiastically commend this book to your attention is noteworthy. Although we have long liked and respected one another, and worked together to advance the work of Resources for the Future (RFF), we had diametrically opposite preferences in the presidential election. That we can agree on the usefulness of this policy manual (without necessarily endorsing every recommendation) is a true confirmation of bipartisanship.

This short volume presents constructive policy options in the form of memos to the president, whether reelected or new to his office, prepared by current and former members of the RFF research staff. We intend these short briefs, constrained to be no more than three or four printed pages, to provide grist for the president and his staff, as well as members of the legislative branch of government, as they seek actionable policies on energy and the environment immediately following the election.

Although this volume contains specific policy options made by scholars associated with RFF, it does *not* constitute the "official" views of the organization itself. Unlike a business trade association or an environmental advocacy group, RFF takes no institutional position on any legislative or regulatory policy matter. Rather, we employ economists and other policy analysts who want to do careful and high-quality research in the fields of energy, environment, and natural resources.

Of course, that does not restrain our research staff from drawing conclusions or offering detailed recommendations. In fact, they have been encouraged to do so in these pages, as they have done on many occasions when testifying before Congress or other deliberative groups. On those occasions, though, they have made clear that they are speaking for themselves and not for the organization that employs them. We recognize that this distinction may seem a fine one to many readers, but it is a defining principle to those of us associated with RFF.

A final observation: the memos in this book were prepared by members of the RFF staff, who were asked, "Based upon your own research and your knowledge of the work of others, what policy recommendation would you like to make to the next president?" We made it clear to our colleagues that they should not confine their recommendations to what the prevailing wisdom says is politically possible. Readers of this anthology will see ample evidence that they followed this advice!

Robert E. Grady
Chair, RFF Board of Directors

Frank E. Loy
Vice Chair, RFF Board of Directors

About the Contributors

H. SPENCER BANZHAF explores nonmarket valuation of air quality and other public goods and has proposed an approach to incorporating public goods into cost-of-living indexes, such as the U.S. consumer price index. He also studies the history of economic ideas and institutions, such as inflation and the agencies measuring it.

THOMAS C. BEIERLE, a former RFF fellow, has looked closely at stakeholder involvement in environmental decisionmaking and has extensively studied the impact of public participation in environmental policy formulation. He currently is an associate with Ross and Associates Environmental Consulting in Seattle.

JAMES BOYD centers his research on law and regulatory economics, including liability law, water quality regulation, ecological benefit assessment, environmental enforcement, and land use management.

TIMOTHY J. BRENNAN, professor of public policy and economics at the University of Maryland Baltimore County, is a coauthor of *Alternating Currents: Electricity Markets and Public Policy* and a former senior industrial organization and regulation economist for the Council of Economic Advisers.

DALLAS BURTRAW has concentrated his research interests on the restructuring of the electric utility market, the design of environmental regulation, and the costs and benefits to society of such regulation. His recent work focuses on multipollutant policy choices, greenhouse gas emissions, tradable emission permits, and valuation of natural resource improvements.

MAUREEN L. CROPPER is a lead economist in the Research Department of the World Bank and a professor of economics at the University of Maryland. She is a member of the RFF Board of Directors and former president of the Association of Environmental and Resource Economists.

JOEL DARMSTADTER studies energy resources and policy, particularly in their relation to economic development. His professional activities have included congressional testimony and participation in National Academy of Sciences studies.

CAROLYN FISCHER concentrates her research on the design of market-based environmental policies, including the costs and benefits of different options for allocating tradable emissions permits. She has explored policies to reduce emissions of greenhouse gases, improve fuel economy, promote technological advances, and better manage natural resources.

ROBERT W. FRI has served as director of the National Museum of Natural History, president of Resources for the Future, and deputy administrator of both the U.S. Environmental Protection Agency and the Energy Research and Development Administration.

WINSTON HARRINGTON pursues research interests that encompass urban transportation, motor vehicles and air quality, and problems of estimating the costs of environmental policy. He has written or coauthored five books, including *Choosing Environmental Policy: Comparing Instruments and Outcomes in the United States and Europe.*

SANDRA A. HOFFMANN examines the use of regulation and tort law in managing health and environmental risks, currently focusing on food safety, valuation of children's benefits from environmental protection, and tort compensation for nonmonetary loss. She is a coeditor of *Toward Safer Food: Perspectives on Risk and Priority Setting.*

RAYMOND J. KOPP led the first examination of the cost of major U.S. environmental regulations, using an approach that is now widely accepted as state-of-the-art in cost-benefit analysis. He is the coauthor of *Valuing Natural Assets: The Economics of Natural Resource Damage Assessment* and is a member of the U.S. Department of State's Advisory Committee on International Economic Policy.

ALAN J. KRUPNICK analyzes the benefits, costs, and design of air pollution policies and also focuses on valuation of health and ecological improvements. He is former senior economist at the Council of Economic Advisers. He is the author of *Valuing Health Outcomes: Policy Choices and Technical Issues.*

RAMANAN LAXMINARAYAN's work on "resistance economics" uses economic analysis to develop policy responses to such problems as bacterial resistance to antibiotics and pest resistance to pesticides. He has served on expert panels on these issues at the World Health Organization and the Institute of Medicine and is editor of *Battling Resistance to Antibiotics and Pesticides: An Economic Approach.*

MOLLY K. MACAULEY devotes her research interests to space policy and the economics of new technologies. She serves on several National Research Council committees addressing space policy and has testified numerous times before Congress on her work.

RICHARD D. MORGENSTERN focuses on the costs, benefits, and design of environmental policies. His research interests include conventional types of pollution as well as global climate change. He has served in senior policy posts in both the U.S. Environmental Protection Agency and the U.S. Department of State.

RICHARD G. NEWELL concentrates his research and outreach efforts on economic analysis of incentive-based policy and the role of technological change in environmental and natural resource policy. His research applications encompass climate change, energy efficiency, energy technologies, valuation of costs and benefits over time, and fisheries policy.

KAREN L. PALMER explores the environmental and economic consequences of electricity restructuring and studies environmental policies focused on electricity generators. She also researches the economics of recycling and product stewardship. A former visiting economist at the Federal Energy Regulatory Commission, she is coauthor of *Alternating Currents: Electricity Markets and Public Policy.*

IAN PARRY specializes in environmental, transportation, energy, and tax policies. His recent work has analyzed gasoline taxes, fuel economy standards, mass transit subsidies, alcohol taxes, policies to reduce traffic congestion and accidents, environmental tax shifts, the role of technology policy in environmental protection, and the distributional impacts of pollution control.

WILLIAM A. PIZER centers his work on econometrics and public finance. He applies much of this work to the question of how to design and implement policies to reduce the threat of human-induced climate change. He is a senior economist at the National Commission on Energy Policy and served as senior economist at the Council of Economic Advisers.

PAUL R. PORTNEY is president of Resources for the Future and is the author or coauthor of ten books, including *Public Policies for Environmental Protection*. He is former chief economist at the Council on Environmental Quality.

KATHERINE N. PROBST focuses her research on the costs and implementation of the Superfund program. She has also examined the management of long-term risks at sites in the nuclear weapons complex. She is lead author of *Superfund's Future: What Will It Cost?*

ELENA SAFIROVA uses economic modeling to examine urban issues, including the impact of land use and transportation interaction on policy effectiveness. She has also studied the effect of telecommuting on urban spatial structure and conducted cost-benefit and distributional analysis of road pricing.

JAMES N. SANCHIRICO conducts economic analysis of fishery policy design, specifically the effects of transferable fishing quotas and marine protected areas. This research is based on ecological-economic models and has received numerous awards. He also works on the dynamics between land use and water quality and biodiversity conservation.

ROGER A. SEDJO directs forest economics policy research, including global environmental problems, climate change and biodiversity, public lands, international forest sustainability, timber supply and trade, forest biotechnology, and land use change. He has written or edited 14 books related to forestry and natural resources.

LEONARD SHABMAN centers his research on studying market incentives in environmental management, including water supply and quality, flood hazard management, river restoration, fishery management, and public investment analysis. He is former director of the Virginia Water Resources Research Center, and a professor of agricultural and applied economics at Virginia Polytechnic University.

JHIH-SHYANG SHIH engages in quantitative analysis of environmental management and resource policy. He models air quality, risk, surface water, and solid waste management, and studies costs of environmental protection, technology adoption, and renewable energy.

MICHAEL R. TAYLOR analyzes and seeks ways to improve U.S. policies and programs that affect agriculture and food security in Africa, and he works with a consortium of universities to develop analytical and decision tools for risk-based food safety priority setting. Taylor, an attorney, was administrator of the U.S. Department of Agriculture's Food Safety and Inspection Service and deputy commissioner for policy at the Food and Drug Administration.

MARGARET WALLS researches solid waste and recycling, urban land use issues, and transportation. Her work on waste and recycling includes analysis of "product stewardship" programs and the cost-effectiveness of alternative policies. In the land use area, she is currently analyzing transferable development rights programs for preserving open space.

KRIS WERNSTEDT explores policy responses to contaminated brownfields properties in the United States, including innovations in site reuse and voluntary cleanup efforts. His most recent work examines the use of environmental insurance at contaminated properties and the relative attractiveness of different incentives to encourage their redevelopment.

Introduction

*By Richard D. Morgenstern
and Paul R. Portney*

Environmental protection has brought huge benefits to the United States. The country's air and water are cleaner than they were a generation ago. Hazardous wastes are better managed. Land once poisoned by toxic dumps is being reclaimed. These achievements, combined with continuing efforts to preserve and enhance natural resources, contribute daily to Americans' health and their improving quality of life.

But these successes have not come cheaply. The country is now spending in the neighborhood of 2 percent of its GDP—more than $200 billion a year—in predominantly private funds to meet the standards that it has set. Although numerous studies find that, in the aggregate, the benefits of our environmental policies greatly exceed these costs, prudent stewardship requires the people who govern the United States to ask whether we are spending these outlays as effectively as possible—and whether we are getting the greatest possible environmental benefits for our money. We must also ask, of course, whether new problems have arisen that require new solutions.

In those instances where change is warranted, the reasons frequently seem technical. But it is in the nature of politics that the technicians themselves often are not able to make the necessary changes. That requires the attention and persuasive powers of people at the highest levels of government. And in the political atmosphere prevailing among policymakers today, that is an intimidating task. A *Washington Post* editorial, commenting in early 2003 on the policy climate, observes that the environmental debate "has lately become ludicrously, almost hysterically, polarized. . . . With surprisingly few exceptions, both environmentalists and industry lobbyists have convinced themselves that the environmental debate is a zero-sum game. Every new regulation is calculated in dollars lost by one side; every failure is calculated in dead birds by the other side."

This book is intended to build bridges between the specialists knowledgeable about these programs and the policymakers themselves—including the president—who must lead any successful drive to improve environmental management. The beginning of a new administration, whether a president's first or second, is always a time for taking stock, reexamining current practices and prevailing ideas, and looking for new opportunities. In that large sector of federal responsibilities that concerns resources and the environment, many opportunities are available. This book presents some of the more prominent examples.

As a country and as a government, we have learned a lot during the last several decades about what works well in environmental management and what does not. It is clear that some present practices are obsolete, measured by current experience. Why? Circumstances change. New problems arise. Legislation that once was entirely satisfactory now often falls short, perhaps because of new scientific knowledge or new patterns in the regulated industries. Sometimes we have learned better ways to organize environmental programs, but old laws and regulations can hinder and even prevent improvement. And too often, the perfect becomes the enemy of the good when crafting legislation or regulations.

In the coming months, we suspect the country will increase its focus on energy and environmental issues. Concerning the former, there are reasons to believe that the rising energy prices witnessed over the last several years may be less transitory than in the past. The rapid economic growth of China and India, coupled with renewed growth in the western economic democracies, have added to global energy demand at a time when energy supplies have not kept pace. Instabilities in the Middle East, Venezuela, and Africa suggest that we should not be at all compla-cent about future supplies of crude oil, one of the most important energy forms on which we depend.

As regards environmental issues, the regulation of air pollution from coal-fired power plants, the need for mandatory federal action to reduce U.S. emissions of carbon dioxide and other so-called greenhouse gases, and federal policies toward the nation's forests, parks, and other public lands all rank high as candidates for serious policy reviews. Similarly, attention may well focus on the fuel economy standards applicable to new vehicles.

We should acknowledge that some of the proposals put forward in this collection would require increased government spending at a time when the budget deficit makes it difficult to contemplate new programs. At the same time, at least one rec-ommendation would reduce the deficit by increasing government tax revenues. A number of the memos suggest approaches for environmental protection, natural resource enhancement, or energy security, where tighter standards could be justi-fied. Others suggest opportunities to either accomplish more for the money the country is spending now or achieve the current level of protection for less. No country, however wealthy, can afford to pass up these kinds of bargains. Approaches like those proposed for fisheries management, fuel economy stan-dards, outer space policy, and auto insurance reform, for example, likely appeal to Republicans and Democrats alike.

Another caveat pertains to the comprehensiveness of the collection. In one sense, it covers a broad range of issues. The memos relate to programs overseen by multiple federal agencies, are pertinent to broad geographic areas, and would affect a large part of the U.S. economy. In all these respects, the issues they cover could be expected to have an impact on almost every citizen in one way or another. Never-theless, it was never our intention to cover every environmental and energy topic—

simply too many exist for us to have been able to prepare a memo on each. Thus, to take but a few examples, we say nothing about endangered species policy, the disposal of radioactive wastes, drinking-water regulation, or water availability in the American West. Astute readers also will note that our focus in this volume is domestic energy and environmental policy, even though many of the most serious problems in these areas fall increasingly outside the United States, most often in developing countries. Notwithstanding these omissions, a number of the memos overlap both in their scope and their recommendations. For example, it would make little sense to adopt both an economywide carbon tax *and* a modified version of the McCain–Lieberman Climate Stewardship Act, which would impose a cap-and-trade system on domestic carbon emissions.

Some of the policy recommendations are sweeping in scope and likely to be quite controversial, such as those advocating a major reorientation of air pollution control efforts toward greater control of particulates, or calling for the imposition of a carbon tax. Others deal with issues that are seldom front-page news. In some cases—public policy regarding antibiotic resistance, for example—this may be because an issue has yet to hit the headlines despite its obvious and growing importance. In other cases, such as the rehabilitation of abandoned properties known as "brownfields," the issue may seem esoteric despite the aesthetic and financial blight they can pose to communities and their inhabitants.

The president will find no recommendation here for a wholesale reorganization of the federal government to improve energy policymaking, or for a thorough overhaul of the multiple and often mutually inconsistent statutes under which the U.S. Environmental Protection Agency operates. Neither is there a recommendation to require presidential appointees to have significant experience in the areas they are selected to lead, though that is to be desired. We suspect that political determination ought to be left to others, because recommendations like these would trigger so much political upheaval and institutional angst that even the hardiest souls would shy away from them.

How, then, can we think of the memos presented here? As an environmental and energy policy agenda for an administration, they can be approached in several different ways. A few represent novel and even daring policy approaches; others are extensions or revisions to ideas that have been discussed previously in policy circles.

Although all the ideas presented in this volume command support in some segment of society, no claim is made about their universal appeal. For some, the adversely affected groups are well defined and confined to a small number of economic sectors, such as commercial fishermen or NASA contractors. In contrast, other proposals—most notably, those designed to mitigate climate change—affect a much wider swath of society, although in some cases the authors have made efforts to limit negative impacts. Because of the preliminary nature of many of the policy ideas contained in this volume—and the fact that for many of them, the costs depend on critical details that still need to be worked out—we have not asked the authors to put specific price tags on their proposals. Nonetheless, the careful reader will quickly gain an understanding of the relevant magnitudes involved, either on or off budget.

Key Themes

We present the policy proposals in four standard groupings: energy and climate; environment, health, and safety; natural resources; and information decision frameworks. At the same time, several cross-cutting themes are identifiable. The expanded use of market mechanisms and the greater emphasis on economic analysis in environmental decisionmaking are prominent elements of many of the memos. While one might expect as much from the largely economics-oriented RFF staff, the views expressed herein clearly mirror long-term policy trends in the United States and elsewhere. In some respects, the essays in this volume represent the latest installment in those trends.

A second theme of the volume is the call for reform of existing policies and programs. For example, Alan Krupnick's proposal to deemphasize implementation of the ambient standard for ozone in favor of more rapid implementation of the fine-particulate standard is one such reform. In another essay, recognizing the management problems associated with the transboundary movement of air pollution, Krupnick and coauthor Jhih-Shyang Shih argue for greater emphasis on regional as opposed to state-level decisionmaking to implement certain provisions of the Clean Air Act. Additional examples of reform considered in the volume include Roger Sedjo's call to revise and simplify major planning activities at the USDA Forest Service, Molly Macauley's proposal to reform the NASA budget process, and Katherine Probst's recommendation to reexamine spending priorities of the Superfund program. Echoing a theme of caution regarding deregulation, Timothy Brennan argues for a go-slow approach for introducing further competition in retail electricity markets.

A third identifiable theme of the volume is the need for new or more aggressive approaches to enhance environmental and natural resource protection in several important areas. For example, recognizing the growing resistance to antibiotics in the general population, Ramanan Laxminarayan calls for expanded federal efforts, including new incentive-based mechanisms, to reduce nonessential use of these medicines in both humans and animals. Recognizing the potential gains from revitalizing neighborhoods burdened with a legacy of contamination, Kris Wernstedt proposes new federal actions with an emphasis on large-scale, areawide revitalization of multiple properties as opposed to the traditional property-by-property approach to brownfield revitalization. Michael Taylor calls for a presidential initiative to accelerate modernization of the U.S. food safety system. He seeks unification of the responsibility for food safety under a single authority, replacement of outdated inspection programs, and allocation of resources more closely tied to public health risks of food. In another essay, Alan Krupnick and coauthor Sandra Hoffmann suggest broadening the use of performance standards for regulating foodborne pathogens as a means of improving food safety at lower cost.

Drawing on the economic benefits literature, Dallas Burtraw and Karen Palmer advocate stringent and rapid restrictions on emissions of sulfur dioxide and nitrogen oxides from power plants. Important reductions in mercury emissions would be achieved as well. Modest limits on carbon dioxide emissions from the same sources also should begin now as an initial, efficient means of addressing climate change.

Other essays in the volume call for broader, multisector approaches to reducing carbon dioxide and other greenhouse gases.

A fourth theme is that, despite the complexity and contentiousness of the issues, energy and climate change present many opportunities for new policy initiatives. For example, Ian Parry and Joel Darmstadter call for new efforts to limit our dependence on oil. Their recommendations emphasize both the demand and the supply sides of the equation. In another essay, Darmstadter calls for a national "green power" initiative, modeled on renewable portfolio standards now in operation in a number of states. Carolyn Fischer and Paul Portney propose introducing much more broadly tradable credits into the current Corporate Average Fuel Economy (CAFE) standards that govern the mileage requirements for new cars and trucks.

To thwart the possibility of global warming, several proposals call for mandatory, multisector actions. Robert Fri makes a broad-scale recommendation for U.S. leadership on the climate issue. Raymond Kopp, Richard Morgenstern, Richard Newell, and William Pizer advocate five specific revisions to Senators McCain and Lieberman's Climate Protection Act, which was defeated in a 55–43 bipartisan vote in 2003. In another essay, Dallas Burtraw and Paul Portney make the case for a broad-based carbon tax, based on the need both to mitigate greenhouse gases and to address the growing federal budget deficit.

A fifth theme is that better information, along with broader decision criteria, is essential to improving both the effectiveness and efficiency of environmental management. While recognizing that expanded data collection may be costly and potentially inconsistent with the well-established goal of limiting paperwork and related burdens throughout society, a number of authors contend that the gains from better information clearly outweigh the costs. James Boyd and Leonard Shabman, for example, argue that gaps at all levels of government in basic information about water quality are major stumbling blocks to improving the overall quality of our nation's rivers, lakes, and estuaries. They recommend a National Water Quality Monitoring Strategy, developed under federal leadership, and supported by changes to the Clean Water Act, as a means of improving water quality management.

In another essay, Spencer Banzhaf calls for the creation of a new statistical agency—a Bureau of Environmental Statistics—to centralize and coordinate national efforts to develop comprehensive, consistent statistics on the quality of the environment. He proposes to model the new agency after such proven enterprises as the Census Bureau, the Bureau of Labor Statistics, and the Bureau of Economic Analysis. Thomas Beierle's memo recommends efforts to reinvigorate environmental information disclosure as a policy tool. Regarding the criteria used by the regulatory agencies and the Office of Management and Budget for evaluating regulatory options, Maureen Cropper advises that the current focus on benefit–cost analysis be expanded to include greater emphasis on cost-effectiveness as well.

A final theme concerns policy innovation. Just as we all recognize the societal gains from rapid advances in computers, medicine, and other science and technology areas, we need to encourage similar innovation in federal and state policies to improve management of natural resources and the environment. Some of the policies that seemed workable 20 years ago may not be appropriate today, particularly when we acknowledge some of the technological changes that have occurred in the intervening years. Ian Parry and coauthor Winston Harrington, for example, introduce the notion that the global positioning satellite systems being installed on many new vehicles offer the potential to revolutionize the way we pay for our car insurance. Imagine a reliable, efficient system to collect the same revenues that individuals now pay for car insurance, but on a per-mile basis rather than on the current fixed annual charges. As they argue, the potential safety and environmental gains from such a scheme may be considerable, not to mention the ancillary climate change benefits. An essay by Harrington, along with colleagues Karen Palmer and Margaret Walls, proposes a federal "policy auction" to encourage states to introduce innovative policies. The authors provide examples in the fields of waste recycling and the use of market mechanisms to relieve congested roads. In another essay, Parry and Elena Safirova focus on converting some existing high-occupancy vehicle (HOV) lanes into high-occupancy/toll (HOT) lanes. A final innovative proposal by James Sanchirico focuses on U.S. fishery management. To promote both economic growth and biological sustainability, Sanchirico proposes that the new president take aggressive steps to introduce measures to zone the oceans, analogous to the spatial management decisions routinely made regarding land use.

The president who takes office on January 20, 2005 will confront the competing approaches that loom large over questions of energy and the environment: Americans want cleaner air and water and healthy surroundings, but they also want inexpensive fuel, big cars and houses, and economic growth. A president has the platform to encourage changes in personal behavior to promote conservation and responsible stewardship of resources and to move the body politic toward consensus on appropriate policies. Inevitably, the postelection deliberations within and outside of the executive branch will center around options designed to address these competing approaches. We hope that the presidential memos presented here will be thought provoking and, most important, that they will make commonsense contributions to those deliberations.

Part I
Energy and Climate

Taking the Lead on Climate Change

by Robert W. Fri

Mr. President, I urge you to propose and, after wide-ranging debate, to implement a beginning framework for control of U.S. greenhouse gas emissions. This framework, which would include a modest but firm cap on major emissions sources, would help American businesses and consumers by creating a more predictable domestic investment environment. As important, it would provide U.S. leadership in the evolution of international controls on these emissions. Taking this first step need not commit the United States to a longer-range program and could leave time as necessary for further study of the extent and impact of climate change.

The Policymakers' Dilemma

Since ratifying the United Nations Framework Convention on Climate Change in 1992, U.S. policy has largely resisted imposing mandatory controls on domestic greenhouse gas emissions. The rejection of the Kyoto Protocol on the grounds of its adverse economic impact has been the most visible example. This hesitation to impose greenhouse gas controls is understandable, but inaction has its own costs as well. The dilemma for policymakers is how to balance the costs of doing something against the costs of doing nothing.

Dealing with the prospect of climate change is a formidable policy problem, both scientifically and institutionally. Although the basic physics of the effects of increasing concentrations of greenhouse gases is clear, the climate system itself is exceedingly complex. Scientists do not yet understand all of these complexities, nor can they construct computer models of the system to predict with precision the effects of a changing climate. Still, while agreeing that more research is needed, most scientists believe that enough is known now to warrant action to limit greenhouse gas emissions. The problem science creates for policymakers is thus that it is becoming clear that we need to do something before science can tell us exactly what needs to be done.

The scale of the climate issue also creates a daunting problem of governance—the institutional framework within which policy works. A changing climate is global in its effect, which means that no one country can do enough on its own to solve the problem. Even if all countries began to pull together now to contain greenhouse gas emissions, a control regime would have to operate for decades to successfully limit greenhouse gas concentration in the atmosphere. This means that our generation must pass along to succeeding ones the management of the climate issue. The world does not now have institutions that are particularly effective in managing global cooperation, much less on issues that span several generations.

> *A changing climate is global in its effect, which means that no one country can do enough on its own to solve the problem.*

Thus policymakers can find solid reasons in both science and governance for hesitating to impose a greenhouse gas control regime. Setting the right controls is a hard thing to do, and indeed the costs of being wrong could be high. If the necessary level of control turns out to be significantly greater or smaller than estimated, the world could easily spend too little or too much on this problem. Similarly, if the institutions of governance inequitably allocate the burden of controlling greenhouse gas emissions, major emitters such as the United States or China could reap major but undeserved economic gains or face severe losses.

But inaction also is costly, so waiting for scientific certainty or frictionless governance is not necessarily a good policy. Climate change issues are already creating immediate issues for U.S. businesses and consumers, regardless of what long-term policy on climate change may turn out to be. Here are some examples of the real costs that are being incurred right now:

- Businesses producing significant greenhouse gas emissions abroad are in some cases already subject to mandatory controls. For example, the United Kingdom and the European Union have, or are actively seeking to have, caps on greenhouse gas emissions, coupled with trading of emissions allowances among firms. The cap-and-trade policy is widely regarded as a sensible mechanism, as it helps ensure the most efficient allocation of resources. But a U.S. company that finds its least-cost way of meeting a U.K. cap is to reduce emissions at a U.S. plant needs a U.S. policy that lets it take credit for that action. Otherwise the firm is at a competitive disadvantage.
- Electric utilities and other major domestic sources of greenhouse gas emissions make investments that are designed to last for decades. Over that span of time, however, the possibility that government may impose a mandatory greenhouse gas control regime is, though not a certainty, at least a very real risk. Factoring that future risk into today's investment decisions is more difficult without steady guidance from policymakers.
- Other entities becoming involved in the greenhouse issue are creating costs and uncertainty. States are rolling out their own greenhouse gas controls in the form of emissions limits or renewable energy portfolio standards. Insurance companies are debating whether and how to handle climate risks in their policies. Shareholders are proposing resolutions in corporate proxy statements asking for

disclosure of the business risks of potential climate liability. Although the value of these kinds of actions is open to debate, each is a real cost or risk today.

American businesses and consumers must respond to these facts of life whether or not they believe that climate change is occurring. And the absence of a U.S. framework for controlling greenhouse gas emissions makes their responses more uncertain and therefore more costly. Some companies might want to experiment with early action to explore ways of cutting greenhouse gas emissions, but they would be more likely to do so if they knew that they would get credit for their efforts down the road. And companies that find themselves responding to multiple, uncoordinated requirements almost certainly would be able to control costs more efficiently within a single national framework.

Some companies might want to experiment with early action to explore ways of cutting greenhouse gas emissions, but they would be more likely to do so if they knew that they would get credit for their efforts down the road.

Another risk exists in the possibility that the United States will someday subscribe to an international control regime. The costs to the nation of that regime will depend importantly on the details of its design. For example, access by U.S. emitters to low-cost emissions reductions in other countries is essential to minimizing U.S. costs, not to mention the overall cost of the program. Similarly, how the rights to emit greenhouse gases initially are allocated among nations has a major effect on the costs of compliance. Thus, if and when global climate controls become necessary, a regime that at least does not unfairly disadvantage the U.S. economy serves the national interest.

The United States should take a leadership position now in international forums on the design of these controls for two reasons. First, no one else is going to protect the country's interest. Indeed, experience with the Kyoto negotiations and other discussions has already demonstrated that some countries and advocacy groups would like to limit U.S. flexibility in achieving the lowest-cost solution.

The other reason is that policymakers usually seem to make major environmental policy decisions—such as adopting a serious greenhouse gas control regime—very quickly. When that happens, the quality of the policy that goes into place will depend almost entirely on the quality of thinking that has gone before. To illustrate, the negotiation of the Montreal Protocol to control emissions that destroy atmospheric ozone took place quite soon after the establishment of the scientific smoking gun made it clear that something had to be done. That the protocol was a success testifies to the fact that the international community had engaged in more than a decade of discussions about the problem of ozone depletion. Similarly, the sulfur oxide control program in the United States came into being in just a few months because of an exceptional alignment of political interests. But in the background was a long history of both scientific and economic research that enabled its framers to design a complicated program that actually worked.

The odds are good that the climate issue will evolve similarly. Some precipitating event will cause the international community to take serious action, and to do so quickly, before the stimulus to action recedes. And when the flag drops, the time for careful thought will be over; therefore, the policy process up to that point is critical. For the United States not to be a leading player in this process is to risk a control regime that is less efficient and more inimical to our national interests.

Recommendations for Action

This situation calls for a more certain environment in which businesses and consumers can prudently account for the growing pressure for greenhouse gas controls in today's decisions. Absolute certainty is not possible, because the climate issue still involves both scientific and institutional uncertainties. But it is possible to signal the likely direction of policy, thus reducing the risks to more manageable levels.

Accordingly, your administration should propose a policy framework for greenhouse gas controls that addresses three major goals:

- The design and initial implementation of an initial U.S. control regime.
- Forceful advocacy of the U.S. position in forums about the international control regime for greenhouse gases.
- Development of transition rules that, to a reasonable degree, hold harmless early actions to reduce greenhouse gas emissions.

> *Absolute certainty is not possible, because the climate issue still involves both scientific and institutional uncertainties. But it is possible to signal the likely direction of policy, thus reducing the risks to more manageable levels.*

It should be clear in proposing this framework that you intend to open a debate among all stakeholders aimed at reaching a consensus within two years on specific proposals, which would then be offered for implementation. This focus is important to ensure that the actions you recommend are both effective and prudent. Although the content of your proposals will depend on the outcome of this process, the framework should at a minimum include the following elements:

A modest cap-and-trade program for major emitters of all greenhouse gases. This program would provide experience with the functioning of a national market for greenhouse gas allowances, creating a solid foundation for U.S. policy formulation. The Chicago Climate Exchange and trading regimes being implemented in the United Kingdom and the European Union are examples of how such markets are designed and function. These existing markets also reveal some degree of price discovery for greenhouse gas emissions allowances. These data will be useful in establishing an initial cap that allows the trading market to operate effectively without creating unnecessary costs to the participants.

A similarly modest cap on transportation fuels to encourage development of tools for managing upstream control programs would also be desirable. It is unlikely that the cap would affect the fuel price to the consumer very much, but gaining experience with controls on fuels rather than emissions sources is important for the likely design of a full-scale greenhouse gas control program.

Banking of credits for early action would reduce the risks of experimenting with solutions to the emissions problem. It is already clear that companies will want to experiment with various forms of offsets to learn how to make measurable and verifiable emissions reductions outside their own facilities. Sequestering carbon in forests and helping a developing country reduce its emissions are examples of these offsets. It is up to the private sector to try these experiments, and if they are successful, the participants should get credit in any future control regime.

Rules for international offsets earned from participating in emissions reduction programs abroad should be specified. These offsets might arise from trading in non-U.S. allowance markets and from cooperating with developing countries though programs such as the Clean Development Mechanism. In some cases, countries may bank the emissions reductions from these steps for future use.

Technology-forcing programs that go beyond the usual government research and development goals are essential. The Department of Energy is pursuing research programs to create a hydrogen economy and a zero-emissions coal plant. These technologies will limit greenhouse gas emissions over the long term, but policy-makers should also have a technological hedge against the possibility of controls in the somewhat nearer future. For example, gaining operational experience in electric utility service is critical for the successful introduction of integrated gasification combined cycle technology, so the government might want to find a way to encourage the early deployment of a few operational units well in advance of the zero-emissions coal plant. Another possibility would be to replace the current Corporate Average Fuel Economy program based on the efficiency of gasoline use with one based on the level of CO_2 emissions, thus guiding future automotive technology development in a direction that anticipates greenhouse gas controls.

The central issue you need to deal with is balancing the costs of action on climate change against the costs of inaction. The recommendations given here aim to strike that balance by suggesting actions that do not imply a premature commitment to a long-term climate goal or to the substantial costs that meeting such a goal would incur. Indeed, it is not necessary, and perhaps not even wise, to have such a goal in mind in order to reach agreement on what a sensible control framework would look like. Nor is it necessary to incur large costs to gain the experience needed to assure policymakers that reliable tools will be at hand.

> *The central issue is balancing the costs of action on climate change against the costs of inaction.*

What is needed is leadership that reduces uncertainty and increases confidence that however the climate issue develops, our government is actively shaping it with both the long-term national interest and the nearer-term imperative of prudent management in mind. My experience leads me to believe that such leadership would be welcomed by the many U.S. organizations that are already dealing with climate change as part of their daily business.

R.W.F.

2

Stimulating Technology to Slow Climate Change

by Raymond J. Kopp, Richard D. Morgenstern, Richard G. Newell, and William A. Pizer

Sir, in order to effectively address the threat of global climate change, we urge you to adopt a flexible emissions trading program for greenhouse gases, and simultaneously to increase funding for related technology research and development. Such a program will place the United States on a path to address this serious environmental problem without risk to the nation's economy.

The Problem and the Challenge

Climate change is a vexing problem for all nations. Emissions of greenhouse gases, notably carbon dioxide (CO_2), are associated with the combustion of coal, oil, and natural gas used to power much of the global economy. The accumulation of greenhouse gases in the atmosphere threatens to raise global temperatures and cause potentially damaging alterations in the earth's climate.

For the United States and the world, the challenge is to make modest but meaningful changes in the energy system in the near term that set the stage for fundamental change in the long term. These changes therefore must include incentives both to reduce emissions now and, equally importantly, to develop and deploy new technologies necessary to make future reductions. A program involving flexible emissions trading coupled with a climate technology program has the potential to achieve significant results.

The Leading Proposal for Emissions Trading

A convenient starting point for describing such a program is the proposal introduced by Senators McCain and Lieberman in the 108th Congress (the Climate Stewardship Act, S. 139). Modeled after the successful 1990 acid rain emissions trading program, S. 139 would cap CO_2 emissions at year 2000 levels for most large emissions sources beginning in the year 2010. Under such a program, a limited number of emissions allowances are given out or sold, and emissions sources are required to obtain an allowance for each ton they emit. Whereas an ideal tradable permit program administered upstream on fossil fuel production would address 100 percent of national CO_2 emissions, the McCain–Lieberman proposal would cover more than 70 percent, excluding households, farms, and other small sources.

> *A program involving flexible emissions trading coupled with a climate technology program has the potential to achieve significant results.*

Central estimates by the Energy Information Administration, Massachusetts Institute of Technology, and Resources for the Future all suggest that the initial costs of S. 139 would be on the order of 0.05 percent of GDP, rising significantly in the future. Under alternate assumptions, costs could be much higher even in the near term.

In October 2003, S. 139 was defeated by a bipartisan vote of 55–43 with opponents stressing the potential high costs of the measure, along with the economic and environmental consequences of unilateral actions to reduce emissions. Others emphasized that the proposed emissions caps would barely make a dent in global emissions, instead pointing to the potential of technology, including the Bush administration's $1 billion FutureGen program. Finally, the bill lacks a clear vision of where we need to go in the future.

A More Complete Solution

Because it already has a base of support and familiarity among stakeholders, we recommend using S. 139 as a platform for domestic climate policy, but with five crucial changes based on straightforward economics that also address the concerns voiced in the Senate:

- explicitly tie the emissions trading program to a significant increase in federal support for technology research and development;
- include a cost-limiting safety valve to guarantee that the program will not exceed expected costs;
- allow unlimited offsets from approved domestic and international activities;
- specify a long-term emissions path consistent with a stabilization goal, but include a mechanism (a so-called circuit breaker) that makes future, more stringent emissions limits contingent on compliance costs; and
- make any increase in U.S. effort contingent on climate policy progress in both industrialized and developing countries.

Technology R&D

Effectively addressing the climate change problem requires large-scale changes in our energy system over many decades. Therefore, incentives to encourage the development, adoption, and utilization of a fundamentally transformed energy system are of primary importance for sound climate policy.

The federal government has a long history of supporting energy-related technology R&D, based in part on the positive economic spillovers associated with new, more efficient energy technologies and, more recently, on the global environmental benefit of avoiding climate change. However, this support has not kept pace with the increased need for technological solutions, nor has it always been consistent and well spent.

We propose, therefore, that federal expenditures on climate-related energy system technology R&D be significantly expanded to at least $3 billion annually, or about double current levels, and funded via proceeds from the sale of some emissions permits. Recognizing that federal energy R&D has a somewhat checkered past, thanks in part to a large degree of congressional earmarking and funding fluctuations, we propose that the allocation of these funds be subject to a multiyear integrated R&D planning process. An independent commission would have as its explicit purpose the best allocation of R&D funds for long-term cost-effective climate mitigation and would include experts from government, private, and academic circles. The commission would put forward multiyear R&D funding plans, which would be considered by Congress as a package and forward-funded for several years at a time.

Safety Valve

Technology R&D by itself is not enough to effectively reduce greenhouse gas emissions. Rather, a successful domestic policy must create incentives for the private sector to adopt and use available climate-friendly technologies. The price of permits associated with those emissions under an emissions trading system provides this desired incentive.

Technology R&D by itself is not enough to effectively reduce greenhouse gas emissions. Rather, a successful domestic policy must create incentives for the private sector to adopt and use available climate-friendly technologies.

However, the price of permits under such a system should not be allowed to rise arbitrarily high in order to enforce a short-term emissions cap. The economics of global climate change suggest a focus on consistent long-term effort, not short-term targets. Yet a simple cap-and-trade system can face sudden permit shortages as a result of unexpected events, leading to skyrocketing permit prices and expensive efforts to reduce emissions. This happened in the NO_x RECLAIM market during the California energy crisis and in the lead-up to implementation of the NO_x OTC market when program rules and participation were still uncertain.

The simple solution, and one advocated by chairmen of the Council of Economic Advisers under both Presidents Bill Clinton and George W. Bush, is to specify a price at which additional permits will be sold in order to avoid such a shortage—a "safety valve." We recommend that the program start with a very modest price that rises over time—on the order of $30 per ton of carbon in 2010 with an annual increase of 5 percent per year. The price of $30 per ton of carbon works out to be $0.30 per million Btus of natural gas, $0.60 per million Btus of coal, and $0.07 per gallon of gasoline. The choice of a starting price for carbon on the order of $30 per ton is consistent with the range of estimated climate benefits reported in the literature. The 5 percent escalator reflects increases in climate benefits associated with economic growth.

Based on the emissions limits proposed under S. 139, it is almost assured that the permit prices will reach the recommended safety valve. Nonetheless, we believe this is smart policy: Set the target where you would like to go, and set the safety valve at the price you are willing to pay.

Domestic and International Offsets

Two necessary elements of a successful climate change policy are that it achieves its goals at an acceptable cost and that it engages developing countries, points made most forcibly by the Byrd–Hagel Resolution (S. 98, 105th Congress). Offsets address both elements. They allow participants in an emissions trading system the opportunity to purchase additional permits from sources *outside* the emissions trading system that have achieved verified emissions reductions. These offsets might come from other domestic sources, agricultural and forestry projects that absorb and sequester greenhouse gases from the atmosphere, or, most important, developing-country efforts to reduce emissions.

We recommend allowing unlimited verified offsets because it will both lower program costs and engage developing countries that have low-cost opportunities to reduce emissions. Although there are other ways we can and should engage developing countries in climate change mitigation activities, the purchase of offsets from these sources creates the institutional foundation for full emissions trading in the future.

> *Unlimited verified offsets will both lower program costs and engage developing countries that have low-cost opportunities to reduce emissions.*

Long-Term Vision

As signatory to the United Nations Framework Convention on Climate Change (UNFCCC), the United States has an acknowledged long-run climate policy that is meant to lead to the global "stabilization of greenhouse gas concentrations in the atmosphere at a level that would prevent dangerous anthropogenic interference with the climate system."

At this point, it is not known what greenhouse gas concentrations are consistent with this stated policy aim. At the same time, setting a near-term policy without a long-term vision leaves many people wondering why we should do anything. For that reason, we believe it is crucial to place near-term action in the context of a long-term vision of stabilization.

Because of its centrality in the literature, we recommend a 550-parts-per-million global target—a doubling since the start of the industrial revolution. Based on estimated domestic emissions paths consistent with such a target, we translate this into a 2 percent annual decline in the initial 2000-level emissions cap beginning in 2030, so long as the permit price remains below the safety valve. In this way, the cap proceeds as planned if prices are "safe." If it rises above the safety valve, the cap stops declining, and additional permits are sold to keep the price from rising any further.

Appropriate Action by Other Nations

Climate change is a global problem and requires a global response. The domestic proposal we recommend can be effective only if it spurs additional international action. This could come in the form of trading systems or commensurate policies and measures in other countries. Developing countries should gradually move toward mitigation policies, even if part of the cost of these policies is borne by industrialized countries.

> *The domestic proposal we recommend can be effective only if it spurs additional international action.*

We recommend that any increase in U.S. effort beyond 2015—including continued escalation of the safety valve price—be conditional on substantive actions by all of the nations that are major contributors to greenhouse gas emissions and all countries with whom the U.S. has significant trade relations. Currently, many countries, including the European Union, Canada, and Japan, are pursuing domestic actions to meet their obligations under the Kyoto Protocol. Most developing countries have avoided any climate change obligations and policies. Some developing countries, including China, have pursued policies based on other priorities—such as energy security and local air pollution—that have led to reductions of greenhouse gas emissions. All of these measures should be considered in evaluating further U.S. action.

R.J.K.
R.D.M.
R.G.N.
W.A.P.

A Carbon Tax to Reduce the Deficit

by Dallas Burtraw and Paul R. Portney

Mr. President, your administration will need many measures to deal with the federal budget deficit. As part of your deficit reduction strategy, we recommend that you introduce legislation to create a tax on fuels, based on their carbon content, that would increase gradually over time. A carbon tax will lessen the need to raise taxes on labor and capital formation (which we would like to encourage, not discourage); it also will create an incentive to shift to energy sources that contribute less to atmospheric concentrations of carbon dioxide and other pollutants.

Background

The income tax reductions that took place in 2001 helped ensure that the recession that year was mild by historic standards. However, the growth in federal spending for defense, homeland security, entitlements, and discretionary programs, combined with the tax cuts of 2001 and 2003, have left the federal government spending nearly $430 billion more annually than it is taking in. If all of the tax cuts of 2001–2003 are made permanent and spending stays on its current course, the federal deficit will remain stubbornly high for the next several years, even with a healthy economy. After 2008, when the baby boomers begin to retire and draw social security and Medicare benefits, federal spending and the deficit will explode.

We currently are living beyond our means. Unless measures are taken soon to reduce significantly the size of the federal budget deficit, those who are young today (including the men and women fighting for us in Iraq and Afghanistan), as well as future generations, will pay higher taxes, face higher interest rates, and enjoy a less robust economy. These young people may face another burden because of our actions today—they may live in a world that is climatologically, and therefore environmentally and financially, less hospitable than the one we inhabit today. This burden would develop from the accumulation in the atmosphere of carbon dioxide and other so-called greenhouse gases resulting from the combustion of fossil fuels:

coal, natural gas, and petroleum and its by-products. These gases have the potential to trap heat, causing a warmer world.

A Policy Response

Just as losing weight requires eating less, exercising more, or both, deficit reduction requires spending less, taxing more, or doing both. Because of the size of the projected deficits and the powerful constituencies that make both tax increases and expenditure cuts difficult, it is likely that action will have to be taken on both fronts.

More than nine out of every ten dollars the federal government raises are generated by individual income taxes (45 percent), social insurance taxes (40 percent), and corporate income taxes (7 percent). The bulk of the increased revenues needed for deficit reduction will and should come from these three sources. One advantage of broad-based taxes such as these, in contrast to the proposed carbon tax, is that they tend to minimize economic distortions. However, because taxes on labor supply or capital formation discourage activities that should be encouraged, other ways to raise revenue should be considered as well.

Taxing carbon directly is the most efficient way of reducing emissions and will obviate the need for even higher taxes on labor and capital.

One sensible option is a gradually increasing tax on the carbon content of fossil fuels—coal, petroleum, and natural gas. This has the advantage of addressing the burdens we are placing on future generations through the economic and energy choices we are making today, and it would provide increasing revenues in the future when the fiscal burdens of a large elderly population will be the greatest. Like most taxes, a carbon tax will have some indirect adverse effects on labor supply and capital formation. But taxing carbon directly is the most efficient way of reducing emissions and will obviate the need for even higher taxes on labor and capital.

It is very important that a carbon tax be phased in gradually so as not to destroy the value of existing long-lived investments. These include investments in housing and in the motor vehicle stock; people choose where to live and how many and what types of vehicles to buy at least in part based on the cost of commuting to and from work. They also include huge capital investments in industrial plants and equipment (power plants, petroleum refineries, and so on); in the heating and cooling systems in commercial and residential buildings; and in mines, pipelines, electricity transmission and distribution lines, and other parts of the energy infrastructure. If a new tax on the carbon content of fuels were set at too high a rate, it would trigger a sudden reduction in the economic value of some of these investments, disrupt the economy, and be viewed as inequitable.

It is very important that a carbon tax be phased in gradually so as not to destroy the value of existing long-lived investments.

Because the tax on the carbon content of fossil fuels would be phased in gradually, the tax would not yield significant revenues in the short term. For example, a tax of $5 per metric ton of carbon equivalent (mtce) would raise approximately $8 billion in 2006. Nonetheless, the expectation of future revenue increases should have beneficial effects on long-term interest rates and thereby on economic growth even in the near term, both of which will help ease the budget deficit. If the tax increased by $5/mtce every other year, so that it stood at $15/mtce in 2010, it would generate about $26 billion that year, or about 1 percent of expected federal tax revenues. By 2020, however, the tax would stand at $40/mtce and would be generating estimated revenues of $75 billion annually.

A tax on the carbon content of fuels also would have the desirable effect of reducing emissions of carbon dioxide. It would stimulate investments in existing energy-efficient technologies, create economic incentives for the development of new low-carbon technologies, and prompt energy users to shift to less carbon-intensive fuels. This would mean reduced emissions of carbon dioxide, the greenhouse gas of greatest concern, as well as such pollutants as fine particulate matter, sulfur and nitrogen oxides, and hydrocarbons. For instance, the higher gasoline prices that a carbon tax would engender would mean that the higher-mileage "hybrid" passenger cars and SUVs now on the market would gain market share at the expense of vehicles powered by conventional gasoline engines. Similarly, diesel-powered vehicles, which enjoy 30 percent greater fuel economy than their gasoline-powered counterparts, would also gain a market advantage.

In the electric utility sector, a carbon tax would have similar effects. In the near term, it would lead to greater use of natural gas in place of coal for electricity generation. In the long term, it would make wind and other renewables more economically attractive sources of electricity and stimulate the energy conservation industry. Most important, however, it would ignite investment in research, especially on the application of clean coal technology that will enable carbon dioxide emissions from coal to be captured before entering the atmosphere. This is critically important, as coal is the nation's most abundant energy source.

A carbon tax would ignite investment in research, especially on the application of clean coal technology that will enable carbon dioxide emissions from coal to be captured before entering the atmosphere.

It is difficult to say with precision how significant these emissions reductions might be. Nevertheless, a reasonable guess is that by 2014, annual emissions of carbon will be 95 mtce lower than they would be otherwise, a reduction of approximately 5 percent, with nearly half of these reductions coming from the electricity sector. This would represent a significant contribution by the United States toward greenhouse gas mitigation. Similarly, although emissions of sulfur from the electricity sector are capped and unlikely to be affected by a carbon tax, emissions of nitrogen oxides—precursors of acid deposition and ozone, and also a culprit in the premature mortality believed to be associated with fine particles in the air—would fall by 10 percent.

Cautionary Notes

No deficit reduction measure is pain-free, and a tax on the carbon content of fossil fuels is no exception. Both households and energy-intensive industries, especially those that would find it difficult to increase their energy efficiency or shift to less carbon-intensive fuels, would face higher costs. For instance, by 2014, the $25/mtce tax would mean an increase of $0.003 (or about 5 percent) in the average cost of a kilowatt hour (kWh) of electricity generated. Similarly, this tax would increase the retail price of gasoline by about $0.06 per gallon. Finally, the price of natural gas would increase by $0.036 per million Btus used, less than 1 percent of the current delivered cost of gas to electric utilities.

Retail energy prices are significantly higher in Europe than in the United States, because of the much higher energy taxes there. Although prices are lower for European industrial firms than for the retail sector, European firms may face further increases in fuel costs associated with climate policy in Europe. For this reason, U.S. manufacturing firms whose primary competition comes from Europe will not be significantly disadvantaged by a gradually increasing tax on the carbon content of fuels. However, those that must compete with manufacturers in countries closer to inexpensive supplies of natural gas, or in countries where energy is subsidized, will face stiffer competition. Domestic chemical manufacturers in particular may be affected adversely by a carbon tax, as electric utilities bid up the price of the natural gas that chemical producers use as their principal feedstock.

Inevitably, a tax on the carbon content of fuel will pinch those with low incomes disproportionately. Measures exist, however, to ease the burden for those who can least afford the tax, including the Low Income Heating Assistance Program and the Earned Income Tax Credit. Furthermore, low-income households might be even harder hit by other options for reducing the deficit, such as eliminating social programs targeted at them.

> *The longer the time horizon, the more likely it is that those affected by the tax will take measures to avoid it; while this is important from a revenue standpoint, it also means that the environmental benefits will rise over time.*

The revenues a carbon tax will raise depend critically on the opportunities of consumers and commercial and industrial energy users to conserve, improve the efficiency with which they use energy, or shift from higher- to lower-carbon-content fuels. If as many low-cost, or even no-cost, opportunities to conserve energy exist as some in the environmental community suggest, the tax we recommend will raise less money than suggested here. The longer the time horizon, the more likely it is that those affected by the tax will take measures to avoid it; while this is important from a revenue standpoint, it also means that the environmental benefits associated with the tax will rise over time.

Conclusions

The budget deficit will impose difficult choices. Increased taxes on labor supply or on capital formation are usually the most efficient way to raise revenue, but they also will undermine economic growth at a time when growth is a central strategy for deficit reduction. Although a modest and gradually increasing carbon tax will also affect economic growth, it is the least expensive way to reduce emissions.

We recommend a balanced approach that seeks to preserve the incentives for economic growth. As part of a portfolio of budget policies, the tax we recommend will tap a new source of revenue and also reduce pollution by rewarding investments in cleaner technologies.

D.B.
P.R.P.

Slaking Our Thirst for Oil

by Ian Parry and Joel Darmstadter

Mr. President, in response to the dependency of the U.S. economy on a volatile and uncertain world oil market, we urge that you support phasing in a modest tax on all oil consumption. In addition, we believe that you should plan to expand the Strategic Petroleum Reserve (SPR) and actively draw it down in the event of severe and prolonged oil supply disruptions. The oil tax would encourage energy conservation measures throughout the economy, promote R&D on alternative fuels, and help gradually reduce our vulnerability to price volatility over the long term. More active use of the SPR would help cushion the effects of short-term upheavals in the oil market. These two measures are more appropriate for reducing the nation's vulnerability to oil price shocks than are policy interventions to expand domestic oil production.

Nature of the Oil Dependency Problem

The United States currently consumes almost 20 million barrels of oil a day, more than half of which is imported, and the share of imports in U.S. oil consumption is projected to increase steadily over the next 20 years to around 70 percent. This trend raises concerns about U.S. dependency on a world oil market that is increasingly dominated by supplies from the Persian Gulf, where about two-thirds of the world's known oil reserves are located.

Fears that politically unstable Middle Eastern countries have the United States by the jugular have led to many calls for drastically reducing U.S. oil dependence and even achieving self-sufficiency in energy production. Although U.S. oil dependence raises legitimate concerns that warrant a well-crafted policy response, it is important that such a response takes account of several key factors.

For one thing, the oil intensity of gross domestic product (GDP) has declined by about 50 percent over the last three decades with improved energy efficiency and structural changes in the economy, and these trends are projected to continue (see Figure 4-1). This means that a given future oil price shock will cause less economic

Figure 4-1

Trends in Oil Imports

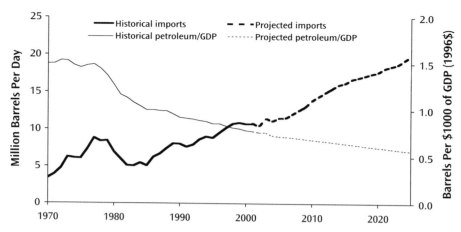

Source: U.S. Department of Energy, Energy Information Administration

disruption, relative to GDP. At the same time, however, anyone who has given serious consideration to the implications of trying to dispense with oil imports altogether recognizes such a goal to be utterly unrealistic—at least for the next decade or so.

Naturally, we would be better off if we could somehow isolate ourselves from the risk of oil price shocks, but reducing oil imports does not automatically reduce our exposure to price volatility. Oil is a fungible commodity, meaning that its price in the United States will be driven by worldwide oil market conditions. The only sure way to reduce the economic disruptions from world oil price shocks is either to reduce the overall oil intensity of GDP through enhanced energy conservation or to lean against oil price volatility itself through more active use of the SPR.

Americans also would be much better off if world oil prices were determined competitively rather than manipulated by Middle Eastern countries acting through

The goal of trying to dispense with oil imports is unrealistic.

OPEC. But again, the ability of the United States to counteract the exercise of market power by OPEC over the long term is limited. Most likely, a reduction in U.S. oil imports would have only a moderate effect on the world price, and it is difficult to reduce oil imports, as opposed to total U.S. oil consumption, or to favor imports from "reliable" suppliers, such as Canada, without running afoul of our international trade obligations. Moreover, a modest reduction in U.S. oil imports may not produce much of a dividend in terms of reduced military spending, given that our Middle Eastern military expenditures serve numerous objectives in addition to oil security, such as attaining peace and stability in the region.

Nonetheless, price shocks still can have substantial macroeconomic effects; recent studies suggest that a 10 percent jump in oil prices reduces GDP by around 0.5 percent, hardly a trivial amount. Is the risk of higher oil prices, and accompanying oil price shocks, likely to increase as production becomes ever more concentrated in the Middle East? There is no sure answer. Although unusually tight market conditions saw mid-year 2004 prices breaching the $45 per barrel level, the long-term demand-supply picture seemed still to point to lower world prices, though perhaps not as low as within the $25-30 price range. One reason for that less alarming prospect might be that it is in OPEC's interest to keep a ceiling on prices. Price jumps could encourage a rapid turn to energy-efficient improvements in oil-importing countries, which would weaken OPEC's market power over the long run. Still, we cannot dismiss the possibility of a political takeover of, say, Saudi Arabia by radical groups determined to harm western economic interests, even though such a move would impoverish their own country.

Given the growing potential for Middle Eastern—and possibly other—governments to harm U.S. economic interests, it seems prudent both to nudge upward the gradual transition to a less oil-intensive economy and to enhance our ability in the near term to respond to oil market disruptions. At least to some extent, the private sector—particularly firms—is aware of the risk of future energy price volatility and already takes into account the benefits of reducing exposure to price volatility when making choices about energy investments, inventory strategy, and conservation measures. But such precautionary steps are unlikely to make the overall economy sufficiently resilient to oil price shocks; for example, firms may not take into account the full costs to society from workers laid off during price disruptions. Hence the case exists for an oil tax and more active use of the SPR.

> *It seems prudent both to nudge upward the gradual transition to a less oil-intensive economy and to enhance our ability in the near term to respond to oil market disruptions.*

The Case for an Oil Tax

We advocate phasing in a tax on all oil uses that, within a few years, would rise to at least $5 per barrel. The $5 per barrel figure is not an arbitrary choice. Recent studies estimate that, by encouraging demand restraint and limiting damage from OPEC exercise of market power, a tax of that magnitude would minimize economic harm to the nation. (This leaves aside environmental and other non-oil-dependence considerations that might independently justify higher taxes.) This tax would strengthen incentives to reduce the use and improve the fuel economy of automobiles, trucks, and aircraft; encourage the substitution of gas and electricity for oil in residential heating; and encourage oil-saving initiatives in petrochemical industries. It would also spur R&D into both energy-saving and alternative-fuel technologies.

A broad oil tax would be much more effective at reducing oil dependence than would an increase in the federal gasoline tax alone or higher fuel economy standards for new passenger vehicles. The broader tax would encourage energy conservation measures and innovation throughout the economy, rather than just in motor vehicles, which account for less than half of the nation's oil uses. A phased-in oil tax of $5 per barrel would ultimately raise gasoline prices by about $0.12 per gallon, so its introduction would not cause any great economic dislocation in the near term; motorists are used to larger fluctuations in fuel prices than this on a year-to-year basis. However, its effect on reducing the oil intensity of GDP, when cumulated over a long period, say 25 years, could be a significant element in overall energy strategy.

An oil tax is a far better approach to reducing our exposure to oil price shocks than supply-side measures, such as tax relief to expand domestic oil production. Increased domestic oil production does not reduce the overall oil intensity of GDP, and therefore it does not reduce the extent of disruptions to energy-intensive activities and production caused by world oil price shocks. Moreover, OPEC might nullify expanded U.S. production by tightening its own quotas.

Even a modest oil tax will have some harmful economic side effects; for example, businesses will suffer from higher transportation costs, and as a result, we would expect some slight detrimental effects on economywide employment and investment. However, a $5 per barrel tax would raise around $30 billion per year in government tax revenues, far more than the annual expenditure required to actively manage the SPR. If, as we recommend, these revenues financed cuts in income taxes, or deficit reduction that effectively reduces burdens on future income taxes, most (though not all) of the harmful side-effects of the oil tax on employment and investment could be offset.

> *Revenues raised from an oil tax could be recycled in income tax reductions or used to lower the country's national debt burden.*

Households would suffer as oil taxes are reflected in higher prices for gasoline and other products; the typical motorist, for example, would pay around $60 more per year in fuel costs once a $5 per barrel oil tax was fully phased in (leaving aside incentives for improved fuel economy). But this does not account for the potential benefit to the country through income tax or national debt reduction, as suggested above. Even then, the oil tax would increase total U.S. taxes on gasoline from around $0.40 per gallon to just over $0.50 per gallon; the hapless motorist in Britain still would be paying seven times that amount.

Use of the Strategic Petroleum Reserve

Oil stockpiling by private corporations is likely to be inadequate from the perspective of the United States as a whole, because firms do not consider macroeconomic risks from oil price volatility when making their inventory decisions. This provides justification for the government to maintain and make use of a strategic petroleum reserve.

The volume of oil in the SPR in mid-2004 amounted to around 660 million barrels, or around 60 days' worth of imports (at its peak, coverage amounted to 115 days in 1985, when imports were less than half their current level). In the past, SPR releases have been governed mainly by federal revenue concerns, rather than as responses to oil price shocks, while the opportunity to purchase oil under conditions of depressed prices was pursued only haltingly.

Although a number of practical issues would need to be worked out, the case for making more active use of the SPR to counteract major oil market disruptions, particularly in conjunction with the other main oil stockpilers—Japan and western Europe—seems compelling. Recent research suggests that the economic benefits from using the SPR to counteract short-term oil supply disruptions, thereby lessening periodic GDP losses and additional oil import costs during a disruption, far outweigh the costs of operating and maintaining the reserve. This research also suggests that expanding the capacity of the SPR to 750 million barrels would yield net economic benefits. Some congressional bills have envisaged an expansion of SPR capacity to its authorized volume of a billion barrels. As world oil market developments and expectations evolve, the rationale for such expansion deserves periodic evaluation.

We emphasize that the SPR should be used only during severe and prolonged price disruptions, and not to counteract year-to-year fluctuations in oil prices. In addition, the more aggressive resort to the SPR that we are recommending does not mean that precise rules governing its release, such as how high and for how long oil prices have to rise to trigger releases, need to be formally spelled out. The particular conditions surrounding a given oil supply disruption will vary. Also, by maintaining a conscious degree of ambiguity about specific trigger points, U.S. policy will make the rewards from deliberate acts of disruption uncertain, perhaps causing those inclined to pursue such acts to have second thoughts about their strategy.

> *The reserves should be used only during severe and prolonged price disruptions, and not to counteract year-to-year fluctuations in oil prices.*

Conclusions

An adequate and affordable supply of energy and resources is a key ingredient in a prosperous U.S. economy, and oil remains one vital component of our energy portfolio. Although we cannot fully insulate ourselves against world oil market upheavals, a modest oil tax and more purposeful use of the strategic petroleum reserve are two important ways of lessening their impact.

I.P.
J.D.

5

Stimulating Renewable Energy
A "Green Power" Initiative

by Joel Darmstadter

Although still a relatively minor player, renewable energy is nearing the point at which, on both technological and commercial grounds, it can begin to play a significantly expanded role in the nation's energy mix. The electric power sector, which currently relies on renewables to generate just 2 percent of its output, is particularly well suited to phase in the use of renewables as a complement to fossil, nuclear, and large hydro resources in the production of electricity. To that end, Mr. President, I recommend that your administration and the new Congress support enactment of a national Renewable Portfolio Standard (RPS). Such a measure, containing explicit safeguards to limit cost increases and providing for an efficient nationwide trading system, would require the use of minimum shares of renewables in electric generation over the coming years. Notwithstanding the failed attempt to adopt such legislation several years ago, an appropriate model to consider in framing the proposed new policy remains the Federal Renewable Portfolio Standard provision (Sec. 606) of the Energy Policy Act of 2002 (S. 1766).

Background

What are the benefits of increased reliance on renewable energy, such as wind, biomass, and other resources? (For our purposes here, "renewables" excludes conventional hydroelectric dams; despite their continuing importance in the Pacific Northwest and elsewhere, an expanded role for hydropower is unlikely.) Several factors come into play. Typically, the use of renewables produces less environmental damage than does using the more conventional energy resources. Notwithstanding emissions restrictions imposed by existing clean-air legislation, a kilowatt-hour (kWh) of electricity generated by coal still causes more pollution than one produced by, say, a wind turbine; and that is before figuring in the problem of greenhouse gas emissions from coal combustion. Then, too, as their name suggests, renewables are relatively immune to the rising cost that, in time, may hit depletable resources like natural gas. Finally, to the extent that renewables replace the use of resources vulnerable to the instability and uncertainty of world energy markets—which, in coming years, may apply not just to oil, but to natural gas as well—their role can make us more "energy secure" and resilient to the effect of internationally triggered energy price shocks.

RPS Programs

How do RPS programs work, and how widespread is their present reach? In principle, their operation is fairly straightforward. A government statute obliges electric utilities to certify that a specified portion of their retail electricity deliveries was generated using nonhydro renewable resources. Typically, the required renewable percentage shares begin at modest levels, gradually rising over a number of years. A key feature of the RPS idea—both in concept and in practice—recognizes that among some utilities and in some service areas, meeting the stipulated renewables percentage may prove more or less difficult. This is where a system of tradable renewable credits enters the picture. Utilities overfulfilling their quota can bank or sell credits representing that excess to firms unable to meet their target. The trading market thus created ensures that the state or, if a federal program, the nation achieves the *overall* mandated target. If, for whatever reason, the cost of renewable credits rises above levels deemed reasonable, price ceilings to shield the program from becoming economically burdensome are an intrinsic safety valve for a workable RPS effort.

To date, only state governments have introduced RPS mandates. (In Europe, several national governments have done so.) Had it passed, the RPS provision of the federal Energy Policy Act of 2002 would have required renewables to account for 2.5 percent of retail electric deliveries in the year 2005 and to rise by annual increments of 0.5 percent, so as to reach a target of 10 percent in 2020.

Among other things, the bill provided for the issuance (by the secretary of energy) of renewable energy credits to producers of electricity generated by wind and other renewable resources; and for the purchase and sale of credits by retail distributors obliged to meet the indicated yearly percentage targets. Importantly, a $0.03/kWh cap guaranteed that the price of credits would be contained. In fact, empirical analysis of a nationwide program and actual experience at the state level suggests credit prices far below such a cap.

What can we learn from state RPS activity? In the last five years, more than a dozen states have launched RPS programs. Although these differ from each other in some respects, the essential features sketched out here are common to all. The Texas initiative can serve as an especially instructive example of a successful program—but also one whose limitation, as in the case of other states, argues for the national program here recommended.

In Texas, currently second only to California in installed windpower capacity in this country, each of the state's utilities is required to deliver 2,000 megawatts (in stages) of renewable generated electricity by 2009, a target that would constitute about 3 percent of the state's electricity supply. Interestingly, the Texas RPS

In the last five years, more than a dozen states have mandated renewable energy standards.

legislation also tightened emissions standards on fossil fuel combustion, underscoring the fact that the market outlook for renewables is strengthened when competing energy systems are held responsible for excessive emissions. Without that provision, the effectiveness of the Texas RPS would have been attenuated by coal's greater competitive advantage.

The Texas RPS regime incorporates a tradable credit system among utilities and generators. An important feature is that it prevents runaway escalation of credit costs through provisions that, in effect, cap credit prices at $0.05/kWh, a level vastly higher than prices at which credit transactions actually have taken place. But the Texas RPS mandate, as in some other states, applies only to trades within the state's electricity network (Some regions pursue a less restrictive policy. Within the Pennsylvania–New Jersey–Maryland, or PJM, interconnection, for example, tradability applies to the multistate region.) The reason is obvious: even if the needed transmission links were available, it would not behoove Texas legislators to have, say, a Corpus Christi utility buy credits from (in other words, reward) an Oklahoma firm rather than one in El Paso. Similarly, to cite another recent example, a Wisconsin utility that has contracted to purchase the output of an Iowa wind farm is able to sell the renewable credits so earned only within Wisconsin.

Although the initiatives taken by Texas and other states provide valuable insights into the benefits of renewable standards, they fail to point up the far greater returns to the nation as a whole from enactment of a comprehensive federal program.

Thus, although the initiatives taken by Texas and other states provide valuable insights into the benefits and challenges associated with an RPS program, they fail to point up the prospectively far greater returns to the nation as a whole that would accrue from enactment of a comprehensive federal program. In contrast to the relatively thin liquidity provided by a statewide market for tradable renewable credits, the vastness of a national market, exploiting the transaction potentials of interregional transmission networks, almost certainly would be economically more efficient. Simply put, a given renewables target could be achieved at least cost. And this does not even make allowance for the saving in what economists view as "transaction costs": the legislative, administrative, and innumerable other costs that each jurisdiction would incur in making its program work. Finally, a less decisive, but perhaps relevant consideration: short of a federal program, some states might seek to compete for the location of electricity-intensive industrial firms by sparing them the prospect of an obligatory renewables element in their electric bills.

Cautionary Notes

Any proposal for a government initiative designed to alter purely market-based outcomes raises several legitimate concerns. A principal one has to do with cost implications. Here, a recent report from the Department of Energy's Energy Information Administration (EIA) provides useful, and reassuring, insights. EIA analyzes the implications of a national RPS target of 10 percent by the year 2020 and a credit cap of $0.015/kWh. The study shows that the RPS requirement, as expected, would lead to greater generation from wind and biomass. Conversely, it would reduce generation from natural gas and coal. The projected nationwide average retail price of electricity would rise by a virtually undetectable 1.5 percent, from $0.067/kWh to $0.068/kWh (expressed in the 2001 price level). Because of reduced demand for the fuel by electric generators, the price of natural gas is projected to drop slightly below levels otherwise prevailing. The price of a traded renewable credit would top out at about $0.01/kWh.

As the EIA's 10 percent scenario conforms to the target of the rejected 2002 Senate bill, that finding has particular cogency here. An independent Resources for the Future (RFF) analysis of a federal RPS policy has produced roughly comparable, and hence equally encouraging, results, though in its more detailed regional dimensions, it also reaches the not surprising conclusion that cost impacts in different parts of the country would vary. Thus electricity costs in the Southeast would rise somewhat more than elsewhere, and they would probably decline in the Pacific Northwest. Additionally, the RFF study finds that with renewables replacing some fossil fuel combustion, carbon dioxide emissions from the electricity sector would diminish by around 5 percent.

Clearly, a driving force in such projections is the expectation that the cost of generating electricity from renewable resources—wind in particular—will continue falling just as it has done in recent years. Still, in spite of that promising outlook, the country's citizens are entitled to ask three relevant questions about this proposal. First, why not rely on strictly voluntary programs to spur greater penetration of renewables? True, a number of such initiatives around the country have offered ratepayers the option of signing up—and paying a surcharge—for the inclusion of a "green" component in their electricity purchases. However laudable, evidence suggests that these programs are unlikely to come anywhere near the magnitude of renewables deployment that even a relatively modest nationwide RPS program would achieve. Moreover, it is questionable whether electric generators, uncertain about the commitment of consumers, would undertake the long-term investments needed to make a voluntary effort more than a marginal complement to an RPS initiative.

Strictly voluntary programs are unlikely to come anywhere near the magnitude of renewables deployment of even a modest nationwide renewable standards program.

Second, even if the federal government can be shown to be a more effective instrument for promoting renewables than the states, why not bypass government altogether and let the marketplace and private sector dictate the use of renewables? But keep in mind that numerous innovative technologies in and outside of the energy field—such as those in aerospace, nuclear science, and information technology—successfully were spurred at least in part by government initiative. To the extent that the social benefits of renewables (cleaner air, a less vulnerable energy system) fall into the category of "public goods" that the private sector would not readily take on, especially early in their development, governmental stimulus can provide an indispensable momentum.

Third, how does an RPS compare with other policy options—some complementary, others alternative—to promote the renewable objective? Clearly, the RPS is not the only public policy avenue to increasing the role of renewables in the electric power sector. Carbon taxes or carbon cap-and-trade programs, for example, could force generators to diminish the use of fossil fuels in their slate of energy sources, giving a boost to renewables (and, conceivably, nuclear power). But the RPS constitutes the most focused federal policy explicitly targeting the use of renewable energy resources in our electricity system. Even production tax credits (PTC), designed to enhance the competitiveness of the particular resource so favored, lack the coherence that the RPS promises. That is because a PTC covering the entire range of renewables surely would turn out to be an intractable and heavy-handed policy to pursue. A fixed across-the-board kilowatt-hour benefit would conflict with the fact that renewable systems vary widely in their stage of development; yet to individually tailor benefit magnitudes would force the government to determine the status of one or another specific technology, even while resisting the blandishments of industries seeking privileged attention in the subsidy queue. All told, an RPS would allow for both a wide variety of renewables options and would give more play to market forces in determining economically efficient outcomes.

To be sure, Mr. President, you will need to address the problem of folding in a national RPS policy under circumstances in which a number of states already have legislated their own, not necessarily compatible, programs (as the proposed 2002 legislation explicitly recognized). But this is not the first or last instance where national and state policies need to be brought into alignment. The ultimate success of a countrywide program also depends on progress in strengthening interconnections among the nation's electric reliability regions. The more robust the interregional linkages, the greater the scope for electricity flows from "renewable-rich" to "renewable-poor" areas.

> *Integrating a national renewables program with existing state programs is not inherently difficult. It will not be the first or last instance where national and state policies need to be brought into alignment.*

Innovative policies inevitably invite at least some skepticism and opposition. Some people point to the intermittency of some renewables: no windpower is produced when the wind is not blowing (at least, not until economic storage technology appears on the scene). This intermittency limits the amount of conventional generating capacity a wind installation can be expected to replace. But at the relatively modest magnitudes contemplated in an RPS regime, we would not normally count on renewables to provide the base-load, "dispatchable" power generated at large power stations.

It is also easy to understand that objections would be raised in "coal-intensive" regions of the country. The trade association representing traditional electricity providers may have had such concerns in mind in stating that a federally mandated RPS "would raise electricity prices for consumers [and] create inequities among states." With respect to the first point, the empirical evidence cited earlier points to electricity prices that seem entirely manageable. As to the specter of wealth transfers among states, it is worth remembering that interstate commerce, enshrined in the U.S. Commerce Clause, is predicated on mutual benefit between buyer and seller—be it a Wyoming coal mine that helps electrify Chicago or an Iowa wind farm that transmits power to Wisconsin.

Conclusions

As in other aspects of energy policy where success depends on both public- and private-sector involvement, a federally initiated RPS program has substantial merit. Mr. President, I encourage you to consider reviving and adopting something similar to the RPS provision of the Energy Policy Act of 2002. Experience with state programs and the EIA and RFF analyses of a national program suggest that the major anxiety about an RPS mandate—a feared escalation of electricity prices—seems largely unfounded. Even if not precisely identical to the 2002 bill, a new federal RPS initiative patterned on that bill would represent a rational and significant step forward in the nation's energy policy. At the same time, it is a modest step that, with the learning experience that often accompanies creative programs and investments, can help lay the groundwork for an even more potent renewables sector in decades to come.

J.D.

Rewarding Automakers for Fuel Economy Improvements

by Carolyn Fischer and Paul R. Portney

S ir, we urge you to introduce legislation directing the National Highway Traffic Safety Administration to write regulations making the "credits" that automakers can earn for exceeding current fuel economy requirements fully salable, both between a given manufacturer's passenger car and light-duty truck fleets and also between different manufacturers. Doing so would provide significant savings to manufacturers and consumers at no cost to the environment. Whereas considerable acrimony surrounds every major effort to change fuel economy standards, the benefits of tradable credits should be uncontroversial, regardless of the prevailing standard.

The CAFE Program

Paralleling current concerns, Congress was worried in 1975 about increasing imports of crude oil, especially from politically and militarily unstable parts of the world. One response was the Energy Policy and Conservation Act of 1975, in which Congress mandated for the first time that passenger cars and so-called light-duty trucks (pickup trucks, minivans, and sport utility vehicles) had to meet fleetwide Corporate Average Fuel Economy (CAFE) standards. Congress itself set the target for passenger cars at 27.5 miles per gallon (mpg), nearly double the pre-1975 average. The National Highway Traffic Safety Administration (NHTSA) was given the responsibility of setting fuel economy targets for light-duty trucks, which now stands at 20.7 mpg—a nearly 50 percent increase over 1975—and is due to increase to 22.2 mpg by 2007.

Working in concert with sharply increasing gasoline prices in the early years of the program, the CAFE standards resulted in significant improvements in fuel economy for both passenger cars and light-duty trucks. As pointed out in a 2002 report by the National Academy of Sciences, some of these fuel economy gains were the result of the rapid downsizing of vehicles, which may have had adverse effects on the safety of these vehicles in crashes, although this opinion was not without dissent. Nevertheless, the CAFE program contributed to significant reductions in oil consumption; as a consequence, between 1977 and 1986, imported oil fell from 47 percent to 27 percent of total oil consumption.

An important feature of the current CAFE program is that it requires each manufacturer separately to meet the standards for each of its own car and light truck fleets. In implementing the program, NHTSA did allow some limited flexibility by taking into account the fact that some automakers might produce higher fleetwide fuel economy in a given year than is required. To reward such performance, NHTSA offers credits for excess compliance and allows the automaker to bank them to offset its own possible future shortfalls against the CAFE standards. However, this offset system currently is limited in several important ways. First, any credits that are earned must be used within three years; after that time the credits expire. Second, credits earned in the passenger car segment of the market cannot be used to offset possible shortfalls in light-duty truck fuel economy, and vice versa. Because most carmakers produce both types of vehicles, this restriction is significant. Third, credits earned by one manufacturer cannot be traded to another. Automakers are allowed to pay a penalty of $5.50 per vehicle for every tenth of a mile that their fleet average falls short of the relevant standard. Domestic manufacturers have never taken advantage of this option, however, always choosing to make and sell enough small vehicles to ensure that they are in compliance with the fleet average requirement.

Fully Tradable Credits

The CAFE program has become very controversial for many reasons, with its critics alleging that it is ineffective, plagued by unintended consequences, likely to result in less safe vehicles, and likely also to spur much less enhanced fuel economy than would result from higher gasoline prices. Although valid responses exist to most of these concerns, they are unlikely to mollify the critics. One change in the CAFE program, however, would work to the benefit of automakers and consumers alike, and should engender relatively little opposition: making the fuel economy credits fully tradable.

To understand the benefits of trade, we should recognize that with an alternate means of compliance, some carmakers might prefer to specialize in the large-vehicle segment of the passenger car or light-duty truck markets because of a comparative advantage they feel they have in manufacturing or marketing such vehicles. They cannot do so now; if an automaker is able to sell 1 million passenger cars that average 26 mpg, it has to sell another million such vehicles averaging 29.2 mpg in order to meet the 27.5 mpg standard. (Although standards are stated in the familiar miles per gallon [mpg], the average is actually calculated based on gallons per mile, which is the rate of fuel consumption.) This has resulted in a situation in which at least some carmakers end up pro-

> *Some carmakers might prefer to specialize in the large-vehicle segment of the passenger car or light-duty truck markets because of a comparative advantage they feel they have in manufacturing or marketing such vehicles. They cannot do so now.*

ducing and selling for little or no profit (or even at a loss) significant numbers of smaller cars or light-duty trucks to enable them to produce the larger cars or trucks on which they make their money.

For this reason, CAFE has not had the same effects on all manufacturers. Some foreign automakers have been able to leverage the popularity of their fuel-efficient small cars and compact SUVs into lower compliance costs for their larger vehicles, allowing them to expand their market share in sedans and light trucks of all sizes. Meanwhile, domestic manufacturers have closely paced their fuel economy gains in line with the CAFE requirements, having to expand production of smaller cars in order to stay in compliance while producing the large vehicles their consumers demand. These disparities result from the fact that each manufacturer must on its own meet the requirement for each of its fleets.

Because what we care about is whether the overall vehicle fleet is meeting the fuel economy goals the nation feels are appropriate, this system makes no sense. If fuel economy credits were fully tradable, an automaker would have another option open to it. If it decided that it could not profitably compete in the small-car (or light-duty truck) market, it could use any fuel economy credits that it had generated in the other segment of the new vehicle market, or it could purchase credits from another automaker that had exceeded its passenger car or light-truck targets in a previous year. Automakers purchasing credits would be those that find it difficult to manufacture and sell enough smaller vehicles to offset their large-vehicle sales. The automakers choosing to sell credits would be those for which exceeding the standard is less expensive than purchasing credits. Both companies would benefit from the exchange.

Eliminating the restrictions on the existing credit banking system and allowing full credit trading would improve the allocation of resources in the auto industry. Automakers would have more freedom to optimize the composition of their fleets, and collectively, they could better allocate their efforts for improving fuel economy. For example, currently, a manufacturer that easily meets the standard receives little or no reward for further improving the fuel economy of any of its vehicles. Meanwhile, a manufacturer struggling to meet the standard as a result of many large-car sales has more incentive than its competitors to improve the fuel economy of both its small and large cars. The same disparity applies with respect to cars and light trucks: if the car standard is more difficult to meet than the truck standard, manufacturers will put more effort into improving car fuel economy than light-truck fuel economy. Yet fuel is fuel, no matter what kind of vehicle burns it. With trading, fuel economy has the same value for all vehicles, so efforts to improve fuel economy are directed wherever they are most cost-effective, whether small or large cars or trucks, imported or domestic.

Fuel is fuel, no matter what kind of vehicle burns it. With trading, fuel economy has the same value for all vehicles, so efforts to improve fuel economy are directed wherever they are most cost-effective, whether small or large cars or trucks, imported or domestic.

The benefit of credit trading would flow through to consumers, as well. The lower manufacturing costs from better specialization and more effective allocation of technologies for fuel economy will translate into lower prices for consumers. Furthermore, the environmental impact would be negligible. The overall fleet of passenger vehicles will have the same average fuel economy. Although some foreign manufacturers have overcomplied with the standards in the past, their average fuel economy has been declining toward the mandate in recent years as they sell more large vehicles, so trading is unlikely to forgo "free" overcompliance. In response to lower prices, some consumers may own more vehicles, but others will buy new vehicles sooner and retire older ones. This will help speed the transition to a more fuel-efficient vehicle stock, leading to better performance with respect to other emissions in the meantime, as newer vehicles are cleaner.

Just as the costs of complying with CAFE increase with the stringency of the standards, so do the benefits of trading. Technical progress has helped ease the costs of achieving greater fuel economy, and emerging technologies will continue to do so as they become more widely available. However, technical progress also can be applied to other vehicle qualities that consumers value, such as power, acceleration, and towing capacity. Trading can help ensure that all qualities—including fuel economy—are applied where they are most valued. These potential savings become more important if we rely on increases in CAFE standards to respond to heightened concerns about oil consumption and dependence. A recent study by the Congressional Budget Office suggests that the costs of tightening CAFE to 31.3 mpg for cars and 24.5 mpg for trucks would be about 17 percent lower if credits were tradable.

Recommendations

We recommend that CAFE credits be made fully tradable, across vehicle types and across manufacturers. Given any fleetwide targets that society demands, allowing trade in fuel economy credits will minimize the costs of meeting these targets. As each manufacturer still has the option to continue as before and meet the standards on its own for each fleet, trading will occur only if it makes both parties better off. Manufacturers and their customers will benefit from lower costs, while the same environmental goals are met.

We recognize that setting fuel economy standards is a contentious issue and must be decided via the appropriate policy process. Our recommendation does not hinge on raising (or lowering) CAFE standards, although trading could allow the United States to raise fuel economy standards to some extent without raising costs beyond what they are today without trading. Nor need we weigh in on the question of separate standards for cars and light trucks to promote trading. Those standards determine the free credits the policy allocates to each vehicle type. That allocation may well be worth debating—whether SUVs should receive more fuel consumption credits than cars, whether all vehicles should receive the same, or whether the allocation should be based on weight and safety factors. But whatever the standards are, we can do better. With tradable credits, we allow the automakers and drivers to decide collectively how best to meet the challenge.

With fully tradable credits, manufacturers and their customers will benefit from lower costs, while the same environmental goals are met.

C.F.
P.R.P.

7

Making Electricity Markets Competitive
How Fast and by Whom?

by Timothy J. Brennan

Expanding competition in historically regulated markets has been a stunning (and largely bipartisan) policy success story over the last 25 years. But thus far, electricity has not been as amenable to similar initiatives. Moreover, recent events, such as the California crisis of 2000–01 and the Northeast blackout in August 2003, have led many to voice concerns about our electricity systems.

We hope that the following provides a useful short guide to help balance the claims these different interests express. In our view, the need to control costs, ensure reliability, and prevent market power in this crucial sector implies that although competition may be politically and economically appealing, presidential leadership should be applied cautiously with regard to opening electricity markets. Successful competition in wholesale electricity markets will require continued regulation and oversight by the Federal Energy Regulatory Commission (FERC). Despite resistance from state regulators and incumbent utilities, this regulation will quite likely require new legislation to require utilities to join regional transmission organizations (RTOs). These RTOs should be empowered not only to coordinate the interstate and international transmission of electricity across the grid, but also to establish enforceable reliability standards. To ensure effective competition, RTO operations must be fully independent from any generation owner; mandatory divestiture of transmission from generation may be the best means to this end.

On the other hand, states should continue to exercise authority over "retail competition," that is, the extent to which households and businesses can choose their own electricity supplier. Local economies bear the costs and reap the benefits of opening retail electricity markets, and each state can learn from others what works and what does not. Moreover, the benefits of extending deregulation beyond large industrial and commercial users to household customers may not be worth the trouble.

Implementing Competition

Opening electricity markets follows a deregulatory trend going back to at least the Carter administration. Economists long had recognized that regulation was handicapped by the difficulty faced by the government in getting the information necessary to set reasonable prices. In addition, the process tended to respond more to the needs of the regulated firms rather than to those of the customers.

Replacing regulation with competition has been very successful in sectors that are workably competitive, most notably airlines, trucking, and banking. In other sectors, introducing competition has been more difficult because segments within those industries retain tendencies toward monopoly. The telecommunications sector, for example, includes competitive markets in services (long-distance telephone, Internet access) and goods (customer telephone equipment, large-scale switching systems). All of these markets, however, depended on access to local telephone service, viewed during the 1970s and 1980s as a monopoly to remain regulated. The antitrust-based divestiture by AT&T of these local monopolies in 1984 was predicated on the belief that the competitive markets required some insulation from regulated monopolies, to ensure nondiscriminatory access and prevent the use of monopoly revenues to engage in inefficient or predatory cross-subsidization.

These concerns strongly influence electricity policy. Generation (the production of electric energy) and marketing (the sale of electricity to final users) have become amenable to competition. Long-distance transmission systems and local distribution grids used to move electricity from generators to users, however, remain regulated monopolies. Opening generation and marketing segments to competition requires some degree of separation from the regulated transmission and distribution segments that typically had been provided by integrated utilities. Introducing competition into electricity has come to be known as "restructuring," reflecting the need to "restructure" utilities to provide at least functional, if not formal, separation between the competitive and regulated markets.

The stakes are high. The electricity sector constitutes about 2 to 3 percent of the gross domestic product in the U.S. economy, comparable to, if not exceeding, the sizes of the automobile manufacturing, agricultural, or television industries. But this 2 to 3 percent figure, large as it is, vastly understates electricity's importance. As the August 14, 2003, Northeast blackout reminded us, our economy and society grind to a halt absent electricity to provide light, refrigeration, heating and cooling, communications, transportation, and the energy to power our factories, businesses, and homes.

> *The stakes are high. The electricity industry constitutes about 2 to 3 percent of the gross domestic product. But that figure, large as it is, vastly understates electricity's importance.*

Numerous groups have stakes in whether and how competition is implemented. Incumbent utilities and new independent power producers are each major players with conflicting motives. Large manufacturing firms want to use competition to obtain electricity on favorable terms and conditions. Because electricity generation is a leading emitter of greenhouse gases, nitrous oxides, sulfur dioxide, and other pollutants, environmental advocates have a strong interest in the sector's performance. Suppliers of generation using "renewable" fuels espouse policies to promote their technologies as a way to cut down national consumption of fossil fuels. Most of all, the general public cares very much about ensuring that electricity supplies continue to be reliable and affordable.

Setting Policy

The intricacy of the decades-long policy puzzles presented by this sector follows from electricity's inherent nature. Besides the fact that it is crucial, electricity possesses two other properties that make it inherently difficult to manage. The first is that unlike virtually all products, at any moment the amount of electricity consumed must just equal the amount produced, neither exceeding nor falling short. Electricity cannot be stored economically for times when demand exceeds supply. Excess supplies of electricity cause lines to overload. Shortages are costly, in that they lead not to minor inconveniences, but to blackouts. Second, because the grid is interconnected, a failure of one supplier to meet demands from its customers leads to a blackout for everyone. This makes the reliability of the electricity system a "public good" that an unregulated market is unlikely to provide adequately.

Because the grid is interconnected, a failure of one supplier to meet demands from its customers leads to a blackout for everyone. This makes the reliability of the electricity system a "public good" that an unregulated market is unlikely to provide adequately.

The intricacy of U.S. regulatory institutions may rival the engineering complexity of the system. Both states and the federal government play significant and legally intertwined roles. Federal authority, exercised primarily by FERC, covers "wholesale" electricity markets—the sale of electricity from generators to load-serving entities (LSEs), primarily local utilities, which provide electricity to end users. FERC's jurisdiction over wholesale markets extends to the transmission system used to carry electricity from generators to the LSEs. "Retail" electricity policies, affecting prices customers pay and whether customers get to pick their LSE, rest in the hands of state legislatures and public utility commissions.

Wholesale and retail markets have separate histories with regard to restructuring initiatives. Wholesale competition began as an unintended by-product of the 1978 Public Utility Regulatory Policy Act (PURPA), passed in response to concerns over energy independence. PURPA's nominal goal, to promote conservation by requiring utilities to purchase electricity from a limited set of approved generators, was notoriously expensive. However, PURPA showed that the grid could work with nonutility generators. Congress enacted the Energy Policy Act in 1992 to open access to the transmission system to any independent power producer wishing to compete in the wholesale market.

Four years later, FERC issued the enabling regulation toward this end, Order 888. Since then, through its Order 2000 (issued in 1999) and more recent proposals to standardize wholesale markets, FERC has encouraged the development of RTOs. Ideally, RTOs would have sufficient breadth to reflect the growing geographic areas over which electricity is bought and sold, and sufficient independence to ensure that access to the grid is open to all suppliers without discrimination. However, energy legislation introduced in the last Congress would have limited FERC's ability to require utilities to join RTOs and to mandate specific wholesale market designs. In addition, the August 2003 outage has raised questions about whether federal authority is sufficient to ensure that the transmission system can limit the frequency of blackouts and prevent them from spreading across the continent.

In contrast to wholesale markets, retail electricity sales are under the control of fifty-one jurisdictions (including the District of Columbia). Roughly half the states have not pursued retail competition at all. The strongest efforts have been in the Northeast and California, states with relatively high electricity prices and thus the strongest sense that competition might improve local economic conditions. When first proposed and put into practice, California's initiative was the most celebrated, implemented in the spring of 1999 following extensive study and publicity and a unanimous vote by its state legislature. At the same time, Congress saw efforts from both sides of the aisle to encourage, if not force, states to open their retail markets.

These efforts in Congress and by the states to open retail markets to competition were slowed, halted, and in some places reversed by the meltdown of the California market in the summer of 2000. Momentum shifted despite reasonable performance in California's retail markets for more than two years before the crisis and retail competition that has worked without major problems in other states. Determination of all the causes of the California crisis and final disposition of how its costs will be borne by consumers, generators, utility stockholders, and California taxpayers remain open controversial matters.

Issues and Options

Wholesale and retail markets present different problems and fall under different jurisdictions in a federalized regulatory system. Hence, the pertinent policy options for the two are quite different.

A first issue at the wholesale level involves the degree to which the RTO needs to be independent from the utility sector—whether so-called functional unbundling suffices, or whether a more radical full divestiture and creation of a stand-alone transmission company is required. At present, RTO membership is voluntary. Utilities should be forced to join and accept the RTO's authority over the design of wholesale markets, management of day-to-day operations, and ability to require investments to expand the grid to alleviate congestion and facilitate competition in one area from generators some distance away. Transmission expansion is a particularly difficult matter. High prices on congested lines desirably encourage shifting the generation of electricity to sites where transmission capacity is available, but allowing high prices for congestion can create perverse incentives to limit capacity.

This leads to the fundamental and crucial issue of reliability. Traditionally, reliability has been promoted through voluntary standards set by regional power pools and the North American Energy Reliability Council (NERC). Voluntary reliability processes, however, may not suffice as we move from an era in which utilities had geographically separate monopolies to one in which utilities compete against each other to sell electricity to LSEs. NERC may continue to play the major role, but independent RTOs or FERC may need to step in.

A third issue is the control of market power, the ability to profit by withholding output in order to raise price. Although the magnitude of the market power problem in electricity remains controversial, it is widely agreed that at peak demand times when electricity capacity is stretched to the limit, individual generation companies, without colluding with others, may find it profitable to cut energy supplies. Additional restructuring to deconcentrate generation markets, or vesting stronger divestiture authority with FERC, may be required, as our antitrust laws do not themselves keep any individual firm from charging what its market will bear. The profitability of withholding will be even greater in some settings, as in California during its crisis, when retail prices were largely fixed, ensuring that the public would continue to demand electricity regardless of its wholesale price.

On the retail side, a threshold issue is whether to expand the federal role over the segment of the sector that has long been left to the states. Were Congress to exercise more authority over retail markets, FERC or any other regulatory body would face numerous options on a multitude of dimensions. One includes the breadth of retail competition—whether to thrust it upon everyone or to emphasize industrial and larger commercial users with a greater stake in the outcome. Reluctant to cast everyone to the market, state regulators have faced difficult choices in setting a "default service" rate not so high as to impede retail uses of electricity, but not so low as to discourage competition from new LSEs. So far, the balance has tilted largely against the entrants, particularly in sales to households. Consumers have shown little interest in choosing new electricity suppliers, partly because regulators have held down the old utilities' prices, but perhaps because they simply do not want to be bothered. Other issues include the extent to which customers should be encouraged or required to install "real-time" meters, so that they have appropriate incentives to reduce consumption during peak periods, when electricity may be upward of 100 times more expensive to produce than off-peak.

Recommendations

With regard to the retail sector, it remains best to leave decisions on opening markets to the states. Efforts to expand the federal role have met considerable political resistance from states and the utilities they regulated. More substantively, uncertainty over how and whether to open retail electricity markets, and the fact that the consequences of retail decisions fall largely within states, justify keeping retail market decisions under state jurisdiction.

> At the wholesale level, I recommend giving regional transmission organizations the geographic scope and independence necessary to match the size of the markets they manage.

Any federal role (or for the states, their continuing role) should include allowing distribution companies to pass through wholesale costs, to prevent the bankruptcies and financial uncertainties that made a bad California situation far worse. Regulators should also consider leaving households and small commercial users under regulation. Efforts to open markets could be focused on those users with enough at stake to make them willing and able to choose among competing providers and adopt "real-time" prices that could vastly improve the performance of the electricity sector.

At the wholesale level, my main recommendation is to give RTOs the geographic scope and independence necessary to match the size of the wholesale markets they manage. Appropriate RTOs may be very large and perhaps international, such as including Canadian utilities in the Northeast. With growing competition, incentives for individual utilities to open grids to competitors and to adopt standards to ensure reliability are likely to be inadequate. RTOs should be able to set mandatory reliability rules and perhaps be fully divested from the generation and marketing arms of member utilities.

The choice of competition over regulation should be a matter not of ideology, but of the degree to which one or the other better serves consumers and the economy. The jury remains out on whether transmission and distribution grids can be managed apart from generation to achieve cost recovery, ensure reliability, encourage short-term supply-and-demand response to congested lines, and provide appropriate incentives to increase generation, transmission capacity, and the use of innovative control systems. Effective counters to the incentive to withhold power at times when demand stretches generation capacity to the limit, especially solutions that do not involve reregulation of wholesale markets, remain elusive. While reasons for optimism exist, policymakers should always keep in mind that electricity's crucial nature and unique characteristics may render it unsuitable for extensive competition.

T.J.B.

Part II
Environment, Health, and Safety

Cleaning Up Power Plant Emissions

by Dallas Burtraw and Karen L. Palmer

Mr. President, it is now widely appreciated that the major environmental public health threat facing the nation is exposure to fine particulates in the atmosphere. We recommend an expeditious reduction in the emissions from electric power generators that contribute to this pollution, preferably through a legislative initiative that streamlines the patchwork of existing and anticipated regulatory policy. Reduction in fine particulates should be the central purpose of a national environmental initiative to reduce emissions of multiple pollutants associated with generating electricity, but it needs to be integrated with a policy for greenhouse gas reductions as well.

Environmental Threats of Electricity Generation

Electricity generation is a major source of the nation's emissions of sulfur dioxide (SO_2) (68 percent) and nitrogen oxides (NO_x) (22 percent). These emissions contribute to the formation of particulates accountable for thousands of premature deaths each year, especially among old, infirm, and very young populations. Exposure to airborne particulates is the most immediate environmental threat facing the public, and it can be rectified to an important degree by rapid reductions in emissions from electricity power plants. In addition, these pollutants contribute to acid rain and visibility impairment, and nitrogen oxides play a part in the formation of ozone. Emissions from other sources, such as industry and motor vehicles, also should be reduced, but the biggest and most immediate bang for the dollar will come by reducing emissions from power plants.

Rarely will the confluence of science, public health, and economics reach such a degree of consensus as exists today on the issue of atmospheric particulates. Although uncertainties still surround this complex problem, it is widely believed that an aggressive approach to reducing these pollutants—especially reductions of SO_2—will yield tens of billions of dollars per year in public health benefits, far exceeding the cost of this program if it is designed well. But for it to be designed well, the particulate problem must be addressed with a policy that addresses all four of the major pollutants associated with power generation.

Emissions from other sources, such as industry and motor vehicles, also should be reduced, but the biggest and most immediate bang for the dollar will come by reducing emissions from power plants.

Electricity generation is also a significant source of the nation's mercury emissions (40 percent), a highly toxic pollutant that can cause ecological damage as well as neurological damage in children. And electricity generation accounts for 39 percent of the nation's carbon dioxide (CO_2) emissions, the most important greenhouse gas associated with climate change. Although neither of these pollutants results in the formation of fine particles, your policy will have to explicitly address the control of all four to be effective. Such a policy will reduce uncertainty for industry and, most likely, decrease long-term compliance costs.

Recent History

The 1990 Clean Air Act Amendments instituted a program of tradable permits for SO_2 emissions, subject to a cap on annual emissions for the electricity industry. Although it was innovative and even controversial at the time, today this program is cited as a major environmental and economic success. Analysis suggests both that the benefits of the SO_2 reductions exceed expectations in 1990 and that costs are lower than anticipated.

NO_x emissions have not been capped previously at the national level. However, beginning in the Northeast and expanding to a large eastern region in 2004, NO_x emissions from power plants are subject to an emissions cap with trading during the summer months, when ozone pollution is a problem.

At the same time, other regulatory time lines and requirements affect the electricity industry. For instance, new source review is a complicated and litigious process that requires power plants that expand or modernize to install state-of-the-art pollution controls. The Clean Air Act Amendments initiated a process that now requires mercury emissions to be regulated under a rule to be issued by March 2005. Other time lines pertaining to broad-based environmental goals also will have a direct effect on the electricity industry. Importantly, however, CO_2 emissions are not regulated.

Several legislative proposals from Republicans and Democrats in the 107th and 108th Congresses would have regulated various combinations of the four pollutants. The remarkable feature of these proposals was their similarity with respect to long-run targets for SO_2, NO_x, and to a lesser extent mercury. The proposals had two distinguishing features, however: their timetables for emissions reductions varied by more than a decade, and differed with respect to the level of regulation of CO_2 emissions. The administration's proposal would not have regulated CO_2.

The differences among the proposals apparently outweighed the similarities, and no legislation has passed. As an alternative, and partly in response to legal pressure under various parts of the Clean Air Act, the administration proposed to achieve emissions reductions through two rulemakings: the Clean Air Interstate Rule (CAIR) for SO_2 and NO_x, and its companion mercury rule.

The administration of George W. Bush deserves credit for initiating the proposed CAIR for SO_2 and NO_x, which is far superior to the status quo. However, SO_2 reductions under the proposed rule come too slowly. SO_2 emissions would be about 9 million tons per year in the absence of further regulation. The CAIR would begin to reduce emissions by the end of this decade, but it would enable emissions to remain above 4.2 million tons past 2020 as a result of the generous use of emissions banking. Most scientists and economists now agree that a substantial further reduction in the SO_2 emissions cap is called for, again making use of emissions trading to keep costs low. The health benefits that would accompany a 78 percent reduction in SO_2 emissions, from the 9-million-ton annual cap imposed by the 1990 Clean Air Act Amendments to a tighter cap of 2 million tons per year, and an accelerated schedule of achieving this target, would more than justify the higher costs.

> *Most scientists and economists now agree that a substantial further reduction in SO_2 emissions would more than justify the costs.*

Meanwhile, the Bush administration's proposed mercury rule emphasizes keeping costs low. This approach makes sense. Significant reductions in mercury emissions would occur at low cost from efforts taken to meet aggressive SO_2 reduction targets. Large investments specific to further reductions in mercury are not justified, yet.

The regulation of SO_2, NO_x, and mercury in the CAIR and companion mercury rule comes in the absence of a policy for CO_2. Coal accounts for 51 percent of the nation's electricity supply and is the major source of all of these types of emissions, and coal would be especially affected by CO_2 regulations because, at the moment, unlike SO_2 and NO_x, there is no readily available way to remove CO_2 from the smokestack. As recent studies by the Energy Information Administration, Environmental Protection Agency, and Resources for the Future show, three-pollutant legislation that excluded CO_2 would precipitate the retrofitting of coal-fired facilities with ambitious controls for the conventional pollutants that cost billions of dollars, but it would have little effect on coal use and CO_2 emissions. However, a strategy that included regulation of all four pollutants would lead to some degree of fuel switching away from coal to natural gas and renewables. It would also affect decisions about investments in new facilities and in research and development of new technologies to reduce CO_2 emissions from coal.

Finally, the CAIR applies additional layers of regulation to a problem that could be addressed best by removing the myriad of unconnected existing rules and uncertain regulatory timetables. The Bush administration tried unsuccessfully to employ a legislative approach. Nonetheless, an approach that addresses the core issues in a simple, comprehensive, and transparent legislative framework still would be preferable to the CAIR.

A Better Approach

A legislative approach to regulating emissions from power plants can provide the most important environmental gains of the decade and also deliver some benefits to the industry by simplifying the planning process. Your agenda should place the highest priority on achieving reductions in fine particles through reductions in emissions of SO_2 and NO_x at the earliest dates feasible. The technology to do so is mature, so the main consideration in setting the date is the time frame necessary to plan for the significant amount of construction activity that would occur at existing plants and for the disruptions that may occur. The opportunity to use an allowance bank to give credit for early emissions reductions at some facilities will help ease the transition to the tighter standard for the rest of the industry. Within a decade, the target for the industry's average annual level of SO_2 emissions should be 2 million tons.

Reducing mercury emissions has received the greatest attention recently. Efforts to reduce SO_2 and NO_x will provide a serendipitous and nearly proportional reduction in mercury emissions. Tightening the SO_2 target to 2 million tons per year will result in a reduction in mercury of about 50 to 70 percent. Furthermore, accelerating the SO_2 reductions also will accelerate mercury reductions. However, special controls aimed specifically at mercury reductions have high and uncertain costs and effectiveness. Targets for mercury reductions should be set to go slightly beyond the level that would be achieved ancillary to SO_2 and NO_x controls, in order to provide an incentive for a modest level of investment that would help reduce the cost of technology specifically aimed at mercury removal from coal plants. But the policy should not require a large-scale application of controls aimed just at mercury, at least for several years into the future, until we learn more about the actual reductions that would be achieved commensurate with further SO_2 and NO_x controls. Meanwhile, policy should look outside the electricity sector to find opportunities for further reducing mercury that can be achieved with a much bigger bang for the buck.

In the long run, the most vexing issue facing the industry is CO_2. Reductions in SO_2 and NO_x can be achieved by policy that does not address CO_2, but paradoxically, this omission would constitute a significant policy with respect to CO_2. The necessary investments to comply with SO_2 and NO_x emissions reductions will extend for decades the operating life of a fleet of coal-fired power plants for which it is technically impractical to incorporate technologies to reduce CO_2 emissions. Hence, your policy with respect to regulating multiple pollutants in the electricity sector will constitute a policy with respect to CO_2 emissions, whether or not you choose to include CO_2.

> *Simple principles for addressing CO_2 provide some guidance. First, it is more important to start soon than to start big.*

Simple principles for addressing CO_2 provide some guidance. First, it is more important to start soon than to start big. Starting soon in requiring CO_2 emissions reductions provides a signal that future constraints are likely, thereby guiding new investments and R&D. Starting soon provides a chance to develop institutions that one day may be significant, and it captures low-hanging fruit in the form of low-cost emissions reductions available now. One day, the nation may agree that it is important to go much farther in the way of CO_2 emissions reductions. In the meantime, it makes sense to take initial steps now. We recommend a slow-stop-reverse strategy. Your administration should announce its intention to cap annual CO_2 emissions from the industry beginning in 2010 at the level of emissions expected for that year. In subsequent years, the CO_2 cap should decline by a small percent each year. (For a complementary proposal, see Chapter 3.)

A second principle is that it is very important for CO_2 regulation to end up as an economywide program; it is not important that it start there. It makes some sense to begin reducing CO_2 emissions in the electricity sector, because this sector is such an important source of relatively low-cost emissions reductions. For meaningful reductions to be achieved, however, the reductions eventually must be economy-wide, so it is critical that the regulatory architecture that takes shape in the electricity sector can naturally be expanded to the entire economy.

In 2005, the central purpose of environmental policy for the electricity sector should be the rapid reduction of particulates, primarily through reduction of SO_2 and NO_x emissions. For this policy to be effective, however, it must integrate policies for the regulation of mercury and CO_2. With such a comprehensive package of environmental regulations for electricity generators, using a cost-effective regulatory institution becomes more important than ever before. A cap-and-trade program provides the mechanism to achieve these reductions in a cost-effective manner.

Cautionary Notes

This policy will have different effects on electricity producers depending on the types of technologies and fuels they employ. Producers who generate primarily with coal will be the most adversely affected by these policies, as they are the primary emitters. Coal plants are likely either to install expensive pollution controls or to be used significantly less. Producers that burn natural gas will be less severely affected, because they do not contribute to SO_2 or mercury emissions, but these plants still will bear costs for controlling NO_x and CO_2. On the other hand, producers relying on nonemitting technologies such as hydropower or nuclear power will face zero compliance costs and are likely to benefit as a result of increases in electricity prices and in the costs of their rivals.

The value of emissions allowances under this program will be far greater than the value of allowances under previous cap-and-trade programs, and it will be dramatically greater if CO_2 is included. A monumental issue is how emissions allowances are distributed initially, because this will have an important influence on the compensation that producers receive, the change in electricity prices facing consumers, and the cost of the program to the economy.

A monumental issue is how emissions allowances are distributed initially, because this will have an important influence on the compensation that producers receive, the change in electricity prices facing consumers, and the cost of the program to the economy.

Free allocation of emissions allowances is the approach used in the SO_2 and NO_x trading programs. But free allocation, especially in the case of CO_2, can overcompensate producers for their costs and places an unfair burden on consumers. To remedy this, a portion of emissions allowances can be auctioned, rather than given away for free, thereby providing compensation that is proportionate to costs and generating revenues that can be used for a variety of potential purposes. The size of the portion to be auctioned is related to the pollutants regulated and emissions reductions achieved under the policy. In the case of a CO_2 policy, the evidence is strong that a major portion of emissions allowances should be auctioned, thereby producing revenues that could be used for compensating fuel suppliers and mining communities.

Allowance revenues also could be used to encourage new renewable technologies or conservation. However, the most efficient use of auction revenues would be to reduce the size of the deficit, and we recommend that this be a top priority. This recommendation comes with the cautionary note that an auction of emissions allowances may lead to higher electricity prices than would other politically convenient but less efficient options.

> *The better approach is a farsighted policy aimed at slowing and ultimately reversing CO_2 emissions in the electricity sector, which would allow for integrated planning to achieve substantial reductions of SO_2 and other pollutants.*

The decision about how to distribute allowances initially cannot avoid an explicit trade-off between economic efficiency and the political goal of keeping electricity prices low. This is a secondary concern for a program that controls just the conventional pollutants, but it is centrally important in a program that controls CO_2, because the efficiency consequences stemming from the way in which allowances are distributed are so large in this case.

Conclusions

Exposure to fine particulates in the atmosphere is the major environmental public health threat facing the nation today. Regulatory uncertainty and the current morass of environmental regulatory requirements and timetables facing electricity generators frustrate industry-planning efforts. The solution to both problems is an expeditious reduction in the SO_2 and NO_x emissions from electric power generators that cause this pollution, which also will yield significant commensurate reductions in mercury. The preferred approach is a legislative initiative that streamlines the patchwork of existing and anticipated regulatory policy. Furthermore, your policy will constitute a policy with respect to controlling CO_2 emissions, whether or not you decide to directly regulate CO_2 emissions. However, the better approach would be to initiate a farsighted policy aimed at slowing and ultimately reversing increases in CO_2 emissions in the electricity sector, which would allow for integrated planning to achieve substantial SO_2, NO_x, and mercury emissions reductions within the next decade.

D.B.
K.L.P.

Pay-As-You-Drive
for Car Insurance

by Winston Harrington and Ian Parry

We urge, Mr. President, that your administration implement measures, particularly federal tax credits, to accelerate experiments in automobile insurance reform. The experiment of most interest is pay-as-you-drive (PAYD) insurance, which varies much more strongly with mileage than most automobile insurance plans currently do. If this policy reform works as its advocates predict, it would reduce the number of highway deaths without increasing overall insurance costs for the average motorist. Environmentalists also embrace PAYD, because reduced driving would moderate other problems associated with motor vehicle use, including oil dependence and emissions of conventional pollutants and greenhouse gases.

Background

More than 40,000 people die in motor vehicle accidents each year, and another 3 million suffer injuries. The costs to society from these fatalities and injuries, and additional costs from property damage, travel holdups, emergency services, and so on, have been measured at several percent of gross domestic product.

Although driving is much safer than in the past, thanks to improvements in vehicle technology (airbags, shatterproof windshields, and so on), better roads, and reduced drunk driving following raises in the legal drinking age, the number of road deaths and injuries is still substantially higher than it need be. One reason is that people do not consider the full societal cost of accident risk when deciding how much and how often to drive. Although they may take into account the risks of injury to themselves and other family members when making these choices, they are unlikely to consider other costs, such as the risk of injury their driving poses for other drivers and pedestrians, the costs of vehicular damage that is covered through insurance claims, and the costs to other motorists held up in traffic congestion caused by accidents.

Automobile insurance is the principal mechanism for coping with these accident externalities. Accident insurance premiums are "experience rated," individually tailored to risk factors known to be associated with causing accidents, such as one's age, sex, residential location, and prior crash record. However, almost universally, premiums are levied on an annual lump-sum basis and therefore vary greatly with the number of vehicles owned by a household, but only vary moderately with how much those vehicles are driven. Consequently, the current insurance system does not induce motorists to fully consider accident risks when deciding how much and how often to drive.

Almost universally, premiums are levied on an annual lump-sum basis and therefore vary greatly with the number of vehicles owned by a household, but only vary moderately with how much those vehicles are driven.

Unfortunately, per-mile insurance has failed to get much attention in the past because of the difficulty of measuring mileage reliably. A decade ago, a related idea arose, motivated primarily by environmental considerations.

In 1994, the EPA convened an advisory committee nicknamed "Car Talk" to attempt to find a consensus on the means of reducing greenhouse gas emissions from motor vehicles. Among the ideas that generated much discussion was a "pay-at-the-pump" (PATP) insurance scheme, wherein a portion of auto insurance would be collected by means of a surcharge (of about $0.25 to $0.50 per gallon) on the price of gasoline; lump-sum annual premiums would be reduced accordingly. This proposal, which traded on the commonsense notion that accident risk is strongly related to miles driven, and therefore gasoline consumption, had two attractive properties: it created an incentive to drive less without raising the cost of owning and operating a vehicle for the average motorist, and it was easy to collect and difficult for motorists to avoid. The proposal would have required each state to impose surcharges or risk losing some federal transportation funding.

However, the proposal was strongly attacked by the insurance industry, on the grounds that the surcharge per gallon would be the same for all drivers, regardless of a driver's age, sex, location, and previous crash record. In its attack, the industry was enunciating an important and sensible principle: if insurance reform is adopted for environmental reasons, it must first make sense from an insurance perspective. The PATP proposal fails this test in significant ways.

The Case for Pay-As-You-Drive Insurance

With PAYD insurance, existing rating factors would be used by insurance companies to determine a driver's per-mile rate, and this rate would be multiplied by annual miles driven to calculate the annual insurance bill. Thus by allowing companies to charge higher per-mile rates to drivers at higher risk of causing accidents, PAYD avoids the major flaw of PATP. PAYD relies on the notion that once individual risk factors have been controlled for, mileage driven is an important risk factor. Although the public literature contains surprisingly little conclusive evidence on this point, what does exist suggests that it is reasonable to assume that total accident risk is proportional to total mileage.

An additional advantage of PAYD is that, unlike PATP, it can be voluntary and therefore introduced gradually over time as people gain familiarity with the idea of mileage-based insurance. Insurance companies can market PAYD insurance alongside their conventional insurance plans. Drivers with low annual mileage would have an incentive to opt for per-mile insurance, as it would enable them to reduce their annual premium payments. Under the current system, both low- and high-mileage drivers pay approximately the same annual premiums, even though the former are less likely to crash and make claims on insurance companies. As low-mileage drivers begin to opt out of lump-sum insurance, over time this would increase rates for the remaining drivers, in turn providing them with more incentive to switch to mileage-based insurance.

Under the current system, both low- and high-mileage drivers pay approximately the same annual premiums, even though the former are less likely to crash and make claims on insurance companies.

We believe that the universal adoption of PAYD insurance would have a striking effect on motor vehicle accidents, perhaps saving up to 5,000 lives per year. For example, a driver between ages 25 and 70 with no recent crash record typically pays about $800 per year in liability and collision insurance premiums for a family car or minivan up to several years old. If the premium was levied on a per-mile rather than lump-sum basis, the per-mile costs for the driver would increase by 6.6 cents for a vehicle driven 12,000 miles per year. This would have the same effect on raising the marginal costs of driving as would increasing the federal gasoline tax from $0.184 to $1.50 per gallon, for a vehicle with fuel efficiency of 20 miles per gallon. The rise in per-mile driving costs would be higher still for drivers in a higher risk class, such as youths with prior crash records.

The political opposition by motorists to a sevenfold increase in the federal gasoline tax would be overwhelming. A very attractive feature of insurance reform is that it can achieve much of the desired effect of higher gasoline taxes without a politically suicidal increase in the tax burden on motorists. This is because the increase in the per-mile cost of driving is offset by the corresponding reduction in annual premiums; the average motorist could still choose to drive the same amount as before the insurance reform and be no worse off. In fact, many motorists—at least those with lower-than-average mileage—would end up paying *less* in driving costs under PAYD than they currently do.

Why hasn't the industry itself introduced PAYD? For one thing, it is no panacea. Only the mileage of the vehicle can be monitored, not who is driving the vehicle. Thus when a youth drives his or her parents' car, the rate per mile does not change, even though the driver is a much bigger statistical risk than either parent. More seriously, until recently, no reliable and accurate way has existed of collecting mileage information from motorists. However, new technology, in the form of Global Positioning Systems (GPS) and nearly tamperproof odometers, has largely eliminated this problem, though in the latter case motorists still would be required to bring in their vehicles for periodic odometer inspections.

More seriously still, a PAYD system has external benefits that cannot be captured by the issuing insurance company. When an insurer charges by the mile, its costs are reduced to the extent that its own customers reduce their accident risk by driving less. However, the costs of other insurance companies also are lowered, by reducing the risk of involvement in a two-car accident for their own customers. These other cost savings cannot be captured by the company offering the mileage insurance. This is a significant market failure, and it potentially justifies policy intervention to jump-start the transition toward PAYD insurance.

Civil libertarians have raised some concerns about privacy issues and the possible misuse of information obtained through mileage monitoring about individuals' driving habits. GPS information, in particular, would allow a motorist's movements to be traced. However, safeguards can be considered if issues actually do emerge, and, as long as PAYD insurance is voluntary, drivers can weigh the potential loss of privacy against the benefits of the policy, similarly to the decision about whether to use cell phones, which also raise privacy concerns.

Recommendations

To provide a jump-start toward using PAYD, we recommend altering the taxation of companies providing auto insurance in favor of per-mile insurance. You could do this by granting the companies an annual federal tax credit of at least $50 for each customer that is charged on a per-mile basis. Revenue costs could be capped by limiting the tax credit to the first, say, 25 percent of policies switched to a per-mile basis, and by including sunset provisions. The policy might be funded by a tax penalty for all remaining customers charged on a lump-sum basis; this penalty would initially be very small, but it would grow over time as more customers switched to mileage insurance, providing an ongoing impetus for the transition to PAYD.

To provide a jump-start, we recommend altering the taxation of companies providing auto insurance in favor of per-mile insurance.

This federal tax initiative would reinforce trends that are just beginning to emerge at the state level. In Oregon, insurance companies have been offered a state tax credit of $100 per motorist for the first 10,000 motorists who sign up for PAYD insurance plans. In 1999 in Texas, the Progressive Insurance Company of Ohio was granted permission to conduct an experiment with PAYD insurance. The company was able to sign up 1,100 customers. The experiment was canceled in 2001, apparently because the rarity and expense of the GPS devices required by the plan kept demand to a minimum. Nonetheless, the Texas legislature subsequently passed new legislation authorizing auto insurance companies to offer per-mile insurance. State governments in Maryland and Connecticut also are considering measures to encourage PAYD.

We believe that tax incentives for mileage-based insurance offer both a highly effective and politically feasible opportunity to effectively reduce the numbers killed on U.S. highways. And as a bonus, it would provide substantial reductions in greenhouse gas emissions and other externalities associated with auto use.

W.H.
I.P.

State Innovation for Environmental Improvements

Experimental Federalism

by Winston Harrington, Karen L. Palmer, and Margaret Walls

O ne of the virtues of the U.S. federal system of government is the ability to allow for geographic differences in policies and experimentation across state lines. But not enough experimentation takes place, because its costs are borne locally while the benefits spread across the country. Mr. President, we urge you to use the power of the federal purse in a "policy auction"—a competition among states and localities for federal funds to implement creative new policies, ones that all other states could learn from and emulate. We particularly urge you to focus the auction on the use of economic incentive-based policy instruments for environmental protection; two applications would be congestion pricing experiments for roads and incentive-based instruments to encourage product steward-ship and recycling. The possibility of doing policy experiments is one of the greatest advantages of a federal system, but despite their value, practical barriers prevent many experiments from being implemented.

The Value of Policy Experiments

Policymaking is often a leap in the dark. Despite the most careful analysis, it is impossible to know for sure the effects or the costs of a policy. Policies often have completely unintended consequences. Uncertainty about outcomes also prevents risk-averse policymakers from giving many potentially useful policies a fair hearing.

In a more perfect world, perhaps, there would be a way for policymakers to implement a policy, observe its effects, and then rewind the clock, allowing a fine-tuned policy to be reimplemented at no cost or inconvenience to anyone. Unfortunately, such do-overs are possible only in children's games and marriage annulments. But the American political system potentially offers the next best thing: a federal structure, in which the central government is supreme but the individual states and local governments still retain significant powers in many policy areas. A federal system permits policies to be tailored to local preferences and conditions, moving government closer to the people, and it allows thoughtful policy experimentation by states or local governments. As Justice Louis Brandeis wrote in a famous Supreme Court opinion 72 years ago, "It is one of the happy incidents of the federal system that a single courageous state may, if its citizens choose, serve as a laboratory; and try novel social and economic experiments without risk to the rest of the country" (*New State Ice* v. *Liebman 1932*). Following are some examples of such experiments.

- *Worker safety.* In 1836, Massachusetts enacted the first child labor law in the United States. Several states followed this example, but it was not until 1938 that Congress enacted legislation that effectively ended child labor in manufacturing.
- *Human rights.* Massachusetts was also the first state to outlaw slavery, which the state Supreme Court ruled was incompatible with the state constitution in 1793. The territory of Wyoming was the first place in the country to extend the right to vote to women; when admitted to the Union in 1890, it became the first state to grant female suffrage.
- *Road finance.* Oregon imposed the first gasoline tax in 1919. By 1932, when the first federal gasoline tax was authorized, all states had imposed such taxes.
- *Environmental regulation.* California implemented the first motor vehicle emissions standards in 1959, anticipating the national government by more than a decade.
- *Land use.* New York City enacted the nation's first comprehensive zoning regulation in 1916. Today few local jurisdictions of any size do not have a zoning ordinance. (Houston, Texas, is the most notable example.)
- *Electricity restructuring.* In 1996, California became the first large state, after New Hampshire, to pass legislation opening retail electricity markets to competition and initiating a transition to full deregulation of the price consumers pay for electric energy. Several states, including Massachusetts and New York, followed suit. After a disastrous summer of high prices and rolling blackouts in 2000, California suspended retail competition in 2001, effectively halting all federal efforts to impose retail competition nationwide. Meanwhile, other states remain optimistic that their approach to retail competition will continue to avoid the problems California encountered.

The Supply of Policy Experiments

Despite the numerous examples of interesting state-initiated policy experiments that we can cite, one thing is certain: there have not been enough of them. When a state undertakes a policy intervention of this sort, its citizens bear the risks of a failed experiment, as California did, yet they capture only some of the potential benefits. The policy produces new knowledge of what works and what does not, and this can be used by other states and the federal government to design further policies. The potential value of this information is large, yet few of the information benefits accrue to the state implementing the policy—and paying the price. Other states are likely to want to free–ride. This "market" failure leads to states underexperimenting with new policies.

Examples of Incentive-Based Policies

One thing is certain about state-initiated policy experiments: there have not been enough of them.

Economists have long advocated the use of pricing mechanisms, either directly by taxes or indirectly by construction of artificial markets in permits that are limited in supply, to solve persistent social or economic problems. One particularly compelling example where this approach might work is the imposition of user fees on congested roadways.

Traffic congestion, an inevitable by-product of modern life, is much worse than it has to be. Urban road access is a scarce, valuable resource, yet it is freely accessible to all drivers. Not surprisingly, demand exceeds supply, inevitably leading to lengthy queues, meaning that motorists "pay" for road access even when it appears to be free. But payment in waiting time is an utter waste, whereas payment in road tolls most assuredly is not. Tolls are a revenue source that can be used to provide travel alternatives such as transit, build new roads, or even reduce other taxes.

A more complete explanation of the virtues of road pricing can be found in Chapter 11. Despite these virtues, true congestion pricing—road tolls high enough to appreciably affect use—is uncommon, not only in the United States, but throughout the world. Most examples of successful road-pricing experiments are either "HOT" (high-occupancy/toll) lane experiments, which open up high-occupancy vehicle (HOV) lanes to single drivers willing to pay a toll, or levies on newly constructed lanes. HOT lane experiments are valuable and instructive. To make a real dent in the urban transportation problem, however, pricing reform is needed for the great majority of existing road capacity that is currently free.

Despite its theoretical promise, motorists, and therefore politicians, evidently remain very leery of road pricing. At least part of this leeriness is fear of the unknown. A few real-world experiments could go far toward either allaying these fears or validating them in the most direct possible way.

Another example is using an incentive policy such as a combined product tax and recycling subsidy to reduce solid waste, increase recycling, and promote product stewardship. Recently, managing household waste has become a greater challenge than in the past. After peaking in the early 1990s, recycling rates for many of the products traditionally collected in curbside recycling programs—aluminum cans, PET plastic containers, and glass bottles—have declined. This decline has taken place at the same time that the percentage of the U.S. population with access to curbside recycling has grown. And while these traditional components of the household waste stream are recycled less, the quickening pace of technological change has contributed to a growing number of new products in the waste stream—obsolete computers, cell phones, and other electronic products. Used electronics pose a concern for disposal because of their sheer volume and the fact that they contain hazardous materials. Addressing waste and recycling problems has always fallen to states and localities in the United States; the federal role is minimal. In light of these issues, the time has come for states to be more creative in how they address solid waste and recycling concerns.

One particularly attractive approach for dealing with many waste problems is a combination of an up-front per-unit tax on products and a back-end subsidy, or refund, for recycling. The refund would provide consumers with an incentive to return products for recycling, and the tax should lead to source reduction. If the tax varies with the weight of the product or, perhaps, with the amount of toxic inputs, it would provide an incentive for manufacturers to decrease product weight or substitute away from toxic chemicals that pose a problem at the time of disposal. Research suggests that a tax-and-subsidy policy of this type is a more cost-effective way to reduce waste and promote recycling than many other alternatives, including a so-called advance recycling fee (ARF), used to pay for collection and recycling programs, such as the one considered in the National Electronics Product Stewardship Initiative. And the tax-and-subsidy approach avoids the illegal disposal problems that might occur with disposal fees and landfill bans.

Where combined tax-and-subsidy schemes have been used, they are typically very effective in promoting recycling. For example, states that have deposit-refund programs for beverage containers—so-called bottle bill states—typically achieve higher recycling rates than do states without such programs. The aluminum can recycling rate in 1999 was 80 percent in bottle bill states but only 46 percent in non–bottle bill states. California and several provinces in western Canada have systems for used motor oil and related products such as filters and containers that rely on an up-front fee charged at time of sale, combined with a return incentive paid to collectors of used oil and oil products. The programs in both locations have been very effective, with recycling rates in excess of 70 percent. Lead-acid car batteries and tires, both of which have deposits and refunds applied to them in many states, also have high recycling rates. On the other hand, product-specific recycling programs that fail to provide incentives have met with limited success. Rechargeable batteries are a good example. These are voluntarily recycled by industry, but no incentives are provided for consumers to return them for recycling. We estimate that recycling rates for rechargeable batteries are approximately 12 to 13 percent.

Despite the benefits of incentive-based programs, many states are reluctant to adopt them because of a lack of political will to increase taxes and potential practical difficulties and administrative costs. The waste problem created by many products—particularly new products such as electronics—is becoming widely recognized, but instead of incentive-based policies, states are instituting or proposing bans on particular products in landfills, ARFs that do not vary with product weight or material content and are not accompanied by a return incentive, and so-called producer responsibility schemes mandating that producers provide collection and recycling services for their products.

Implementing policies that combine up-front product fees with subsidies for recycling has some challenges. High transaction costs associated with administering refunds and sorting returned containers by brand have plagued bottle bill programs in several states. California avoids some of these pitfalls by having all containers returned to redemption centers and not requiring brand sorting. Transaction costs can be lowered even more when the tax-and-subsidy system bypasses final consumers and instead pays the refund to collectors, like the used-oil programs do. Collectors in a system such as this have the incentive to collect as much as possible and would be likely to provide incentives to consumers to return products. Tax-and-subsidy policies also face challenges when the products being targeted are durable ones, such as computer monitors or cell phones, or when products include small amounts of hazardous materials. Policies that are set up and implemented differently in different states and applied to a variety of products could provide valuable information to other states thinking of putting such policies in place. They also could provide information about whether the federal government should play a role in policy coordination.

Policies that are set up and implemented differently in different states and applied to a variety of products could provide valuable information to other states thinking of putting such policies in place. They also could provide information about whether the federal government should play a role in policy coordination.

Encouraging State Experiments with Incentive-Based Policies

The very limited use of incentive policies to combat traffic congestion or to encourage recycling indicates that despite the potential payoff, the perceived risks of failure from implementation of these policies are often too great for any state or local politician to take up. If the federal government can persuade a few states or local governments, or even one, to implement experiments in congestion pricing or the use of combined tax-and-subsidy instruments to promote recycling, the knowledge gained will benefit all states.

However, for some policies, it will take a lot of federal persuasion. We know this is true for congestion pricing, because the federal government has tried to get states to adopt experimental road-pricing programs before, with little result. The Department of Transportation began the Congestion (later Value) Pricing Pilot Program in 1991. This program would provide a local Metropolitan Planning Organization (MPO) up to $25 million in planning grants if the plan led to the actual use of road pricing to reduce congestion. An MPO is the local government entity responsible for preparing transportation plans for the metropolitan area, usually consisting of representatives from each of the area's local governments. The program proved singularly ineffective as a promoter of the use of pricing to ration road access. Although a number of local planning studies led to concrete road-pricing proposals, political opposition to these plans prevented almost all from being implemented.

The federal government has not, to our knowledge, offered financial incentives to states or localities to adopt particular waste and recycling policies. It has, however, put much effort into promoting one incentive-based policy: "pay-as-you-throw" (PAYT) pricing of household solid waste collection and disposal. EPA's Office of Solid Waste has had an active program since the early 1990s, which includes fact sheets, promotional materials for communities, brochures and videos with advice on how to set up PAYT programs and avoid pitfalls and problems such as illegal dumping, and case studies of communities that have implemented such programs. Although waste reductions have been well documented in PAYT communities, only about 4,000 communities nationwide use this form of pricing.

Recommendations

We recommend that the federal government use a policy auction to encourage states to develop and formulate incentive-based policies to alleviate urban traffic congestion and to promote recycling, and to stimulate other policy experiments as well. The policy auction should include the following key elements:

- *A competitive proposal process.* State and local government agencies will be invited to submit proposals for funds to implement actual policy demonstration projects. Because it is difficult to say how much enticement a state or local agency will require to submit a proposal, there should be no limit on the amount requested. Proposals will be judged on the basis of their likely contribution to practical knowledge, their credibility with respect to ultimate implementation, their generalizability to other states or localities, and their budgets.
- *An independent proposal review panel chosen by a committee of qualified experts outside the political process.* With the increase in funding suggested here, some safeguards would be needed to ensure that the program does not degenerate into a pork barrel program. To ensure that the proposals are evaluated on a scientific and not a political basis, they should be evaluated by natural and social scientists knowledgeable about the particular subject area. For example, the Transportation Research Board, as a subsidiary of the National Research Council, is the one organization with the expertise, independence, and prestige to evaluate congestion pricing proposals.
- *No transfers of grant moneys (beyond initial planning grants) before the policy begins to take effect.* As the whole point is to gain actual experience with the incentive policy, and because decisionmakers in the state or local government agencies responsible for implementing a policy will likely face intense pressure to renege, their feet will need to be held to the fire.
- *Ex post facto evaluation by independent researchers.* To get the most out of the experiment and to maximize its credibility, scientific evaluation of outcomes is important. One possibility would be to have the National Academy of Sciences (NAS) convene a group of experts to perform an assessment. NAS has the independence and prestige that would permit a credible assessment; however, the evaluation committee should be unconnected to the committee choosing the applications.

Innovative state policies, such as the use of pricing instruments to reduce congestion or to reduce waste and increase recycling, can be a tough sell. The fact that we have little experience with these instruments in the United States to help inform future policy design is another drawback that contributes to policymakers' reluctance to try them in the future. The program proposed here will use federal money to break through this logjam.

W.H.
K.L.P.
M.W.

Pay as You Slow

Road Pricing to Reduce Traffic Congestion

by Ian Parry and Elena Safirova

Sir, we recommend that you urge Congress to authorize a program that would encourage state and municipal authorities to implement road-pricing mechanisms by removing a variety of legal obstacles to converting high-occupancy vehicle (HOV) lanes into high-occupancy/toll (HOT) lanes and, more generally, by helping pay start-up costs of road-pricing initiatives. Congestion pricing increasingly is being recognized as the only really effective way to alleviate the problem of ever-worsening traffic gridlock in the nation's cities, and its time for implementation has arrived with recent developments in electronic payment technology.

The Case for Congestion Pricing

Traffic congestion imposes substantial costs on society. According to recent studies by the Texas Transportation Institute, travel delays and the resulting extra fuel combustion now cost the nation about $70 billion each year. The average time per year an urban motorist loses to congestion during peak hours has grown from 16 hours in 1982 to 62 hours in 2000. Although detailed methodologies used to compute travel delays and to monetize them are not unanimously accepted by all researchers, those numbers provide an idea of the magnitude of the problem. And congestion is likely to become even worse in upcoming years, with continued growth in vehicle ownership and the demand for driving. Meanwhile, environmental constraints, neighborhood opposition, and budgetary limitations are making it ever more difficult to build new roads to accommodate increasing demand.

Typically, it takes only a modest reduction in the number of drivers to unclog a congested road and get the traffic moving faster. Charging people for driving on busy roads at peak periods is the best way to achieve this; such charges encourage some people to drive earlier or later to avoid the rush hour peak, to take less congested routes into town, to car pool, to use mass transit, or to reduce the number of trips, such as by working at home or by combining several errands into one trip.

Other policy approaches are far less effective at reducing traffic congestion. Increased subsidies for mass transit may help lure some people away from driving on busy roads. But this policy can be partially self-defeating: if roads become less congested at peak period because more people are using transit, this attracts onto the roads some people who were not previously driving at peak period because of high congestion. In short, the roads may just fill up with traffic again; this is not the case under road pricing, however, as the charges discourage people from getting back into the car as congestion falls. The same phenomenon tends to undermine other approaches that do not raise the cost of driving, such as expanding cycle access or promoting telecommuting. And higher fuel taxes, which raise the costs of all driving, whether it is in urban or rural areas or occurs during peak or off-peak periods, are an extremely blunt way to reduce traffic jams; before the recent introduction of road pricing, driving in central London was not much faster than walking, despite gasoline taxes seven times as large as those in the United States.

Forms of Road Pricing

Over the last two decades, a considerable amount of money has been invested in adding high-occupancy vehicle (HOV) lanes to urban freeways to try to induce more carpooling; more than 2,000 lane miles were added at a cost of nearly $9 billion. The results have not been encouraging. Nationwide, the share of carpooling in work trips actually fell between 1990 and 2000, from 13.4 to 11.2 percent, while the share of single-occupant vehicles in work trips increased from 72.7 to 75.7 percent, with slightly declining shares of transit and nonmotorized trips making up the balance.

Nationwide, the share of carpooling in work trips actually fell between 1990 and 2000, from 13.4 to 11.2 percent, while the share of single-occupant vehicles in work trips increased from 72.7 to 75.7 percent.

HOV lanes, at least those currently with traffic flows well below those on parallel, unrestricted lanes, result in underuse of scarce road capacity at peak period. Converting them into HOT lanes by allowing their use by single-occupant vehicles in exchange for a fee while continuing to permit high-occupancy vehicles to use the lanes for free, would benefit many motorists. Those who value the travel time savings enough to pay the fee would benefit, while those who continue to use unpriced lanes may benefit from reduced congestion as some drivers switch to the premium lane. Carpoolers who were already using the HOV lane may be slightly worse off as more vehicles join premium lanes, but ideally the fees would be variable and set at levels to maintain free-flowing traffic, even at the height of rush hour. Tolls would be deducted electronically from accounts linked to transponders as vehicles pass under overhead meters; carpoolers would pass by a manned booth where the vehicle occupancy is briefly checked.

Over time, urban centers could develop a network of linked HOT lanes giving drivers access to any part of the region without major holdups, and allowing local authorities to provide express bus service throughout the region, thereby reducing the need for constructing costly express light rail systems.

To date, only three examples of freeways with HOT lanes exist in the United States: one in Los Angeles, another in San Diego, and a third in Houston. However, serious efforts are under way on HOT lane projects in many other urban centers with severe congestion problems. And many motorists are used to paying tolls that were initially designed to pay for the costs of road construction; it is conceivable that these tolls could be made to vary with the level of congestion.

A number of popular objections to road pricing have been raised; a more detailed discussion of this topic is presented in Chapter 10. But none of those objections really seem to hold much water for HOT lane proposals. One objection is that motorists are opposed to paying for something that they previously used for free. However, under the HOT lane scheme, drivers will not be forced to pay tolls, as they can always use the parallel, unpriced freeway lanes.

For the same reason, it is not true that low-income families, who are least able to afford new taxes, will be driven off the roads; they actually may benefit from reduced congestion on unpriced roads. And, in fact, evidence suggests that it is not the rich who exclusively use premium lanes; in California, people of all income levels use HOT lanes when saving time is important to them. It therefore seems unnecessary to give discounts for low-income drivers using HOT lanes (as required by the recent House bill H.R. 3550), as that undermines the effectiveness of HOT lanes in providing free-flowing traffic.

> *It is not true that low-income families, who are least able to afford new taxes, will be driven off the roads; they actually may benefit from reduced congestion on unpriced roads.*

Another impediment is simply unfamiliarity with the concept of road pricing and skepticism about its effectiveness. For this reason, it makes sense to introduce pricing incrementally, increasing the number of priced lane segments as their success in alleviating congestion becomes evident to the general public. California's two HOT lanes, which have been operating for several years, have demonstrated the ability of variable electronic pricing to maintain free-flowing traffic, and surveys in California now show widespread public acceptance of the HOT lane concept. And despite many predictions that it was doomed to fail, the introduction of road pricing in central London has reduced congestion delays by around 30 percent.

Other forms of road pricing exist as well. Outside of the United States, a number of area (cordon) pricing schemes have successfully reduced congestion in city centers, including London, Rome, Trondheim in Norway, and Singapore. Area pricing makes sense when, as in many old European cities, the central business district is dominated by a maze of narrow, winding streets, making pricing of individual roads impractical. In contrast, congestion in many U.S. cities is concentrated on wide highways and large arterial roads feeding into the center, and in this case, pricing of individual highways and arterials makes more sense. But area pricing still could play a useful role in areas such as Manhattan, where downtown streets are severely congested and people can get around by transit or walking if they are unwilling to pay area fees.

Other schemes might involve pricing of particular elements of the transportation infrastructure, including time-varying tolls on bridges. Such tolls currently exist on bridges over the Hudson River, connecting New Jersey and New York City, and on bridges in Lee County, Florida. Other applications might include fees for access to national parks that currently are congested during peak visiting hours.

These alternative-pricing schemes likely would be met with more political opposition than HOT lane conversions, as motorists are left with no option but to pay if they wish to keep using the same roads. They also are more vulnerable to the criticism that poor people might be forced off the roads. Nonetheless, toll revenues might be used in ways to help the poor, such as by spending it on projects to extend transit access to low-income neighborhoods.

Recommendations

Although decisions about urban road pricing are ultimately the responsibility of metropolitan planning organizations, any of a number of initiatives could be taken at the federal level to jump-start its implementation.

■ *Remove the ban on the imposition of tolls on interstate highways.* Most road networks in metropolitan areas include portions of the interstate highway system, and to the extent that these roads are congested, they need to be priced if a locality is to deal effectively with regionwide traffic jams. Although in principle the Transportation Equity Act for the 21st Century (TEA-21) allows tolling on highway segments as part of the pilot program, to date no state has successfully applied for this authority. The pilot program requires the local authority to show that funds from the state's apportionment and allocations would never be sufficient to pay for maintenance and improvements of the road in question over time, which is very difficult to demonstrate. Alternatively, states can impose tolls on interstate highways if they pay back the federal government for funds already invested in the highway in question. The TEA-21 should be amended to allow local tolling on interstates, without these highly cumbersome restrictions.

■ *Allow all HOV lanes to be converted to HOT lanes.* Again, provisions in TEA-21 prevent localities from converting HOV lanes to HOT lanes unless those lanes are in the Value Pricing Pilot Program (VPPP); most HOV lanes currently are excluded from the VPPP, as it is limited to only 15 congestion pricing projects in the entire nation. The TEA-21 should be amended to permit local authorities to convert any HOV lanes to HOT lanes, whether or not they are currently covered by the VPPP.

■ *Federal aid for start-up costs of road-pricing initiatives.* Introducing pricing on existing roads involves various set-up costs, including costs of installing monitoring technologies and barriers to separate priced and unpriced lanes. And the creation of HOT lane networks in metropolitan areas would require the construction of many additional lanes to link up the existing fragmented systems of HOV lanes, implying a substantial amount of new investment (adding one lane mile costs on average $4 million in right-of-way purchase, labor, and material costs). Toll revenues could cover many of these costs; a study by the Reason Foundation finds that about two-thirds of the necessary investment costs for constructing fully integrated HOT lane networks for the nation's eight most crowded urban centers could be funded by using toll revenues to finance tax-exempt bonds. But extra funding through the federal aid transportation program, such as the VPPP, could help to kick-start road-pricing initiatives.

■ *Introduce legislation to address privacy issues.* Another impediment to the implementation of congestion pricing is the uncertain legal basis for electronic toll collection. Many drivers will be reluctant to have transponders in their vehicles until specific legislation has been enacted establishing for what purposes information collected on driving habits can and cannot be used. This concern will become more pressing with increasing use of Global Positioning Systems (GPS) to make electronic payments.

- *Establish a national standard for electronic-tolling technology.* At present, different technologies exist for electronic tolling, including the EZ-Pass, which dominates northeastern states, and the Fastrak, established on the West Coast. Incompatibility between these two systems imposes an additional burden on long-distance road users, particularly trucks, and for a segment of tourist travelers, as vehicles need to be fitted with more than one transponder, and different transponders might interfere with each other. Legislation to establish a national standard for electronic-tolling technology, before road pricing becomes more prevalent in U.S. cities, would help avoid unnecessary duplication of technology installation costs and facilitate a nationally integrated pricing network.

If the government were to adopt these types of initiatives, we might at last begin to reverse the trend of ever-increasing urban congestion.

I.P.
E.S.

Focus on Particulates More Than Smog

by Alan J. Krupnick

With the costs of meeting air quality standards among the highest of all federal regulations, I recommend that your administration take specific actions to focus more of the nation's efforts on meeting the standard on fine particulates, which is the pollutant that has the most severe effect on public health. Correspondingly, you should reduce, through changes in regulation, the current emphasis on meeting the standard for ozone, which is less potent a pollutant, is too subject to the vagaries of weather, and will be reduced as an ancillary benefit to fine-particulate reductions and as vehicles get cleaner.

The Problem

Throughout the 1990s, with the exception of ozone, most areas were either attaining the National Ambient Air Quality Standards or making good progress toward attainment. With many areas of the country out of compliance for ozone, this pollutant had taken center stage as *the* air pollution problem for the nation.

In 1997, after a contentious debate that ultimately was settled by the Supreme Court in 2003, the U.S. Environmental Protection Agency (EPA) set the first standards on fine particulates (particles 2.5 microns in diameter and smaller, abbreviated as $PM_{2.5}$) and moved to an even tighter standard for ambient ozone, using an eight-hour average measure rather than a one-hour daily peak measure. But in setting these standards, the Clinton administration made a major decision to stretch out the implementation dates for the fine-particulate standards, seeking earlier attainment of the new ozone standard. Some of this difference can be explained by the need to improve the particulate-monitoring network. But mostly this decision was made because of concern over the cost of fine-particulate control and an inadequate appreciation within the administration of its dangers.

Currently, although EPA and states and localities are still acting as if ozone is the main problem, they are beginning to recognize that $PM_{2.5}$ also has to be addressed.

These fine particulates—it is far from clear exactly which—clearly cause serious health effects. Study after study has found strong statistical associations between changes in fine-particulate concentrations and premature deaths, hospitalizations, incidences of chronic respiratory disease, and acute health effects. Ozone, on the other hand, rarely is found to be associated with mortality, and when it is, its potency is less than that of fine particulates. One recent review of the literature found a potency for particulate matter related to mortality over ten times that for ozone. In addition, one of the main constituents of fine particulates, sulfates (which come from sulfur dioxide, or SO_2, emissions), causes visibility degradation. Ozone is most strongly associated with asthma attacks, but even here, it is no more potent that fine particulates. In some studies, ozone is more strongly associated with hospitalization, but it is hard to separate out the effects of different pollutants in these studies.

> *One recent review of the literature found a potency for particulate matter related to mortality over ten times that for ozone.*

In addition, meeting ozone standards is particularly uncertain because of the effect of weather variability on ozone formation. Hot days mean more ozone, so even the most industrious, committed local government may find its efforts to meet the standard frustrated, especially because compliance is based on air quality on the "worst" days rather than a typical average day. Particulates, on the other hand, are less sensitive to weather, and meeting the annual fine-particulate standard depends on average performance.

Ozone also is more likely than fine particulates to be reduced by long-term changes in the economy, such as turnover of the vehicle fleet. Such turnover will reduce the two precursor emissions for ozone, nitrogen oxides (NO_x) and volatile organic compounds (VOCs). The NO_x reductions also will lead to reductions in fine-particulate concentrations, but for most of the country (with the exception of the western states), these benefits will be minuscule. Thus the ozone problem may well be resolved by these long-term factors, leaving fine-particulate reductions as the main challenge.

Finally, a host of federal regulations are slated to address NO_x and SO_2 emissions, the latter being particularly important in fine-particulate formation in the eastern half of the United States. These regulations will help meet both ozone and fine-particulate standards, but full realization of the benefits will take time—perhaps too much time.

Overall, focusing on meeting the standard for ozone drains attention and resources away from reducing concentrations of the more important pollutant—fine particulates. Whereas controls on NO_x generally lead to reductions in both pollutants, efficient control of fine particulates involves quite different strategies. Yet these fine-particulate strategies currently receive less attention and scrutiny than they should.

EPA studies of the costs and benefits of reducing particulates to meet the standard show a cost–benefit ratio in the range of ten or twenty to one, roughly $100 billion of benefits for every one microgram-per-cubic-meter reduction in fine-particulate concentrations. In contrast, the benefits and costs of reducing ozone are about equal; net benefits are zero for reductions in ozone beyond the baseline of the old standard. Thus it makes sense to go after particulate matter and pay less attention to further reducing ozone.

Because air quality standards and their implementation have been among the most contentious of all environmental issues and have effects throughout the manufacturing and energy sectors, as well as on the public, all stakeholders in the Clean Air Act debate would care deeply about this issue. Environmental groups would not concede that efforts should be reduced to meet one standard in favor of efforts to meet another. They want activity on all fronts. Industry positions are likely to be heterogeneous. VOCs, an ozone precursor, are primarily a vehicular pollutant, as are diesel emissions (a fine-particulate component). SO_2, a major fine-particulate precursor in the eastern United States, is a pollutant emitted by industry, electric utilities, and vehicles. NO_x is emitted whenever fuel is burned. Some sectors would find it cheaper to reduce ozone precursors, and others fine-particulate components or precursors.

The position of the states is harder to gauge. Efforts to align transportation and air quality plans in order to meet the ozone standard have been tremendously contentious, expensive, and frustrating for states. In part, this is because precursor reductions, which the states have some control over, do not necessarily translate into ozone reductions because of the high dependence of ozone formation on weather. On these grounds, the states might welcome less focus on this pollutant. At the same time, there is a great deal of familiarity with this problem, whereas fine-particulate reductions are a new problem with unknown challenges.

Obstacles to Progress

The EPA has not been asleep at the wheel. In 1998, an EPA committee, jointly chaired by experts from EPA and Resources for the Future (RFF), led stakeholders to consider implementation issues associated with the new ozone and fine-particulate standards. In particular, the idea of developing integrated attainment plans for these two pollutants was discussed, but little came of it. EPA is still writing the separate implementation rules.

The problem may not be completely subject to administrative discretion. The Clean Air Act focuses only on the degree to which an area is not attaining the National Ambient Air Quality Standards (NAAQS) when designing implementation strategies and deadlines. And even here, there is a clash between severity and practicality. One might expect that areas most out of attainment would be forced to act the quickest to bring down their air quality. But deadlines have been set in the opposite way, for practical reasons. In any event, the severity of the effects of the pollutants and the future path of emissions in the absence of further regulation have not been considered in prioritizing efforts to meet the standards.

The specific issue raised here has not been enjoined in Congress or anywhere else, for that matter. Thus it is difficult to say for certain if a strong constituency exists for or against this proposal. What is clear is that significant ongoing reluctance has existed in Congress and several administrations to reopen the Clean Air Act, which last was amended in 1990. This reluctance was exhibited over the stalled bills for reducing SO_2, NO_x, and mercury emissions, called "three pollutant initiatives," and the Bush administration's own bill (Clear Skies) and subsequent decision to implement reductions in these pollutants through regulations instead.

Recommendations

At least two complementary activities must be undertaken. The first is to educate the public in a balanced way about the relative benefits and difficulties of reducing fine particulates and ozone. EPA's pronouncements about the benefits of its proposed and ongoing programs sometimes tend to lump ozone and fine particulates together, as if their severity cannot be distinguished and their reduction had equal urgency. The issues should include the future time path of ozone and fine-particulate precursor emissions in the absence of further regulation, the relative health and other benefits of controlling each pollutant, and the difficulty of reducing a peak concentration of a highly weather-dependent pollutant, such as ozone.

The second important activity would be to put more research effort into identifying the particulate sizes and types that are most damaging to human health. The fact is that NO_x is a tiny contributing factor to fine particulates in the eastern United States, and evidence that its fine-particulate form (as a nitrate) affects health at all is scant. On the other hand, evidence exists that sulfates, the fine-particulate form of SO_2 emissions, are a key agent in causing deaths from exposure to fine particulates. In general, what is not said about the health benefits of EPA's programs is that *all* the deaths in EPA cost–benefit studies are attributed to fine particulates. Part of the reluctance of policymakers, scientists, and others to make a full commitment to rapidly meeting the fine-particulate standard may be the lack of certainty about which types of emissions to regulate.

With these activities in place, several regulatory and legislative options can bring about this change in emphasis. One regulatory option is to change the deadlines for meeting the new standards. Under rules being promulgated, each area violating the new eight-hour ozone standard (except four areas in California) has to meet the ozone standard in 2007, 2009, or 2010, depending on how severe its ozone problem is. These deadlines could be pushed back further, up to five years, particularly for the less severe areas with short deadlines. At the same time, the fine-particulate deadline, currently 2014 (10 years after designation), could be moved forward in time. These deadlines originally were determined in an interagency process that involved compromises among various agencies in the Clinton administration, and they lie outside the purview of the Clean Air Act. With the Supreme Court invalidation of the EPA plan, the act sets the deadlines. Nevertheless, EPA could move the effective deadlines forward by resisting requests for extensions.

> *Each area violating the new eight-hour ozone standard (except four areas in California) has to meet the ozone standard in 2007, 2009, or 2010. These deadlines could be pushed back further. The fine-particulate deadline, currently 2014 (10 years after designation), could be moved forward in time.*

The timing of the Bush administration's proposed Interstate Air Quality Rule (IAQR), which has a full implementation deadline of 2015, also could be changed. Even with the 2014 deadline for meeting the fine-particulate NAAQS, this initiative reaches its goals too late to be of major help to cities. Pushing the deadline further back just makes this situation worse. The implication is that the IAQR implementation deadline needs to be dramatically moved up in time—more in line with some of the legislative proposals for pollution cap-and-trade systems. Other regulatory efforts affecting fine-particulate concentrations also could be moved up in promulgation, implementation, and deadlines.

Another regulatory option would be to change conformity rules. These are rules set by EPA that govern how areas violating the NAAQS for ozone must align their transportation plans with their efforts to meet the ozone standard. Conformity does not currently apply to $PM_{2.5}$, but it will by the end of 2005. Thus one option is to relax or eliminate this rule for ozone.

A fourth option would be to relax state implementation plan (SIP) planning requirements for meeting the ozone standard. Currently the process is extremely complicated and cumbersome, including the conduct of extensive and expensive computer modeling to demonstrate attainment in the deadline year, plans and demonstrations that "reasonable further progress" in cutting emissions is being made every year, showings that specific emissions-cutting technologies mandated by the Clean Air Act are being implemented, and so forth. Once states are designated to meet the fine-particulate standard, a new process will need to be established for demonstrating attainment and moving toward this goal. By relaxing requirements for ozone reductions at the same time, local efforts in planning, modeling, and implementation can be focused where the payoff is greatest. This relaxation also could apply to transition rules for areas violating the old one-hour ozone standard. Such areas are now supposed to maintain all the mandatory measures they had put in place to meet that standard, as well as reduce emissions by a certain percentage each year.

Conclusions

All air pollutants regulated under the Clean Air Act are not created equal—in terms of their potency, their sensitivity to weather, and their abatement costs—and thus they should not command equal attention. Because so many areas currently violate ozone standards, this pollutant has commanded our attention for more than 20 years. But, the effect of weather on ozone and its relative lack of potency, as well as long-term trends toward cleaner vehicles, particularly in lowering VOCs (ozone precursors), argue for reducing attention on ozone. At the same time, a variety of reasons exist for focusing attention on reducing the key precursors and components to fine particulates, primarily SO_2 emissions and probably diesel emissions. Society's scarce resources and the all-too-limited capacities and budgets of local governments argue for more targeted messages from the federal government and greater emphasis on meeting the most important environmental and health threats. For air quality, this focus should be on the fine-particulate standard, with a deemphasis on the ozone standard.

A.J.K.

A New Approach to
Air Quality Management

by Alan J. Krupnick and Jhih-Shyang Shih

Mr. President, we recommend altering the geographic approach to managing air quality in this country to match the realities of long-range transport of air pollution. It makes little sense to have localities responsible for attaining air quality standards when on average about 75 percent of the problem is not of their own making. Either this responsibility should shift to new regional air quality management institutions, backed by EPA, with ultimate responsibilities still lodged within the states, or it should be lodged at the federal level, with a dramatic reduction in state responsibilities. We recommend that a presidential commission be set up to consider these two quite different options for improving the management of our nation's air quality.

Background

Although air quality generally has improved over the last three decades, some areas, including a big portion of the eastern United States, have a poor record on attainment, and the number of counties in nonattainment has increased with EPA's new tighter ambient air quality standards. For ozone, 474 counties now violate the new standards, up from 125 under the old standards. EPA has yet to issue its final list of counties violating the new fine-particulate standards, but it clearly will add to this list.

To meet these standards, the federal government has been issuing a number of important new national and regional regulations, including the proposed Interstate Air Quality Rule, which mandates an even tighter sulfur dioxide (SO_2) and nitrogen oxide (NO_x) cap-and-trade program; the NO_x SIP Call; the NO_x Trading Program; and several rules that reduce particulate and other emissions from off-road and diesel engines.

Even so, it is unlikely that these rules will be enough to meet the standards in some nonattainment areas, which raises the questions of how best to decide on and implement further control measures. In particular, continuing to follow the Clean Air Act's state implementation plan (SIP) process has some fundamental problems. One of the main ones is that the process is local while the air quality problem is regional.

The SIP process provides each state with the responsibility for developing an EPA-approved plan for reaching the National Ambient Air Quality Standard (NAAQS) by the mandated deadline. For most areas, this involves developing an emissions inventory, implementing a host of emissions reductions measures to apply at the local level, and conducting air quality modeling to show that the measures are enough to meet the standards.

This process is likely to be inefficient to meet stringent air quality standards, because scientific evidence shows that air pollutants, such as ozone, fine particulates, and their precursors, can be transported long distances across state boundaries. For example, recent research at RFF and Georgia Tech has shown that, on average, local emissions account for only 23 percent of local ozone and fine-particulate concentrations (Table 13-1). Thus even if one state reduced its ozone, ozone and its precursors still would be transported across its borders from other states. Because the pollution control efforts of upwind states affect the air quality in the downwind states, one state cannot tackle the ozone problem by itself; it will take multiple states' efforts to solve the problem.

> *Recent research has shown that, on average, local emissions account for only 23 percent of local ozone and fine-particulate concentrations.*

In spite of these findings, under the current regulatory system and SIP design, an individual nonattainment state is required to submit only its own SIP for approval; hence the upwind state is not likely to account for the beneficial impact of its pollution control policy on its downwind neighbor, nor the fact that it may be cheaper to reduce emissions in an upwind area than in the area itself. However, cap-and-trade programs can "automatically" address this problem to some extent.

Previous Attempts to Address the Issue

Regional approaches to solve air quality problems are not new. The Clean Air Act (CAA) provides states with the authority to sue for problems of "overwhelming transport" and gives EPA the authority to set up organizations to address air pollution problems spanning multiple states or tribes. Specifically, petitions under Section 126 of the CAA, the formation of the Ozone Transport Assessment Group (OTAG), and the creation of the Ozone Transport Commission (OTC) are good examples of attempts—albeit incomplete ones—to address this issue.

Lawsuits are a clumsy way to implement policy, however, so this approach should not be relied on to reach attainment. Lawsuits cannot address how states will meet their own attainment problems or those of the region on a cost-effective basis.

OTAG was a partnership among EPA, the Environmental Council of the States, and various industry and environmental groups, which assessed the long-range transport of ozone and its precursors. Although it was successful as a means of information exchange among the states, it lacked regulatory authority and failed to recommend measures to address the problem.

Finally, OTC, a multistate organization comprising government leaders and environmental officials from 12 northeastern and mid-Atlantic states, has developed cap-and-trade strategies to help states in the region attain and maintain the

| Receptor state | 1-hour ozone | 8-hour ozone | Episode average 24-hour $PM_{2.5}$ | |
	Reduced point-source NO_x	Reduced point-source NO_x	Reduced point-source NO_x	Reduced point- and area-source SO_2
AL	36	34	35	36
DE + MD	21	20	12	14
GA	24	23	24	27
IL	13	10	11	19
IN	19	17	18	22
KY	25	25	22	23
MA + CT + RI	16	20	19	21
MI	12	13	16	15
MO	12	14	21	12
NC	22	21	19	22
NJ	21	21	10	15
NY	14	13	16	20
OH	15	15	6	15
PA	16	15	17	21
SC	32	32	33	42
TN	41	39	38	35
VA	24	24	21	16
WI	67	67	69	40
WV	17	17	10	25
Average	23	23	22	23

Table 13-1

Summary of State Contributions (%) to Local Air Quality Per Unit Emissions Reduction (July Episode, Area-weighted)

Note: Table shows local contributions to reductions in one-hour and eight-hour daily maximum ozone from unit reductions in point-source NO_x, and reductions in 24-hour $PM_{2.5}$ concentrations from unit reductions in point-source NO_x and reductions in point- and area-source SO_2 emissions. For example, if each state reduced point-source NO_x emissions by one ton, then 36 percent of the resulting ozone reductions in Alabama would be attributed to reductions of NO_x within Alabama. Source: Shih et al. 2004. Source-Receptor Relationships for Ozone and Fine Particulates in the Eastern U.S. RFF Discussion Paper 04-25. Washington, DC: Resources for the Future.

NAAQS for ozone, and do so cost-effectively. However, recent research has shown that the states covered by OTC are mostly the areas violating air quality standards, in other words, downwind states. By limiting the universe of allowable trades within the Ozone Transport Region, the trading restrictions make it difficult for sources to find trading partners; the smaller the trading region, the smaller the potential for lowering the cost of meeting a given cap.

Recommendations

A couple of very different approaches are possible for dealing with the long-range transport issues. You could build regional air management partnerships (RAMPs) via expansion of existing regional institutions or establishment of new ones, or you could eliminate most parts of the SIP process—except for assuring that local emissions do not increase—and shift responsibility for attainment primarily to the federal government.

Building Regional Air Management Partnerships (RAMPS)

Economists generally argue that the geographic reach of institutions regulating environmental pollution should be coincident with the geographic reach of the pollution. In the case of ozone and fine particulates, such an institution would need to be composed of all the states and tribes in an airshed working collectively to address these regional air pollution problems. A collective process could provide the states an opportunity to jointly consider the issues and to explore strategies that are not only cost-effective, but also equitable for every state within the entire airshed.

> *We recommend establishing a new institutional mechanism to address the regional air quality management issue.*

To pursue this option, we recommend establishing a new institutional mechanism to address the regional air quality management issue. The RAMPs concept is consistent with the recommendation made by the Subcommittee for Ozone, Particulate Matter, and Regional Haze Implementation Programs and in a 2004 National Research Council report, *Air Quality Management in the United States*. RAMPs could be implemented by either expanding the current OTC or creating a new regional organization to address the long-range transport problem.

The CAA gives EPA authority to establish transport commissions and air quality control regions. EPA would be able to establish RAMPs using this statutory authority, relying on its general rulemaking authority to provide direction and schedules. RAMPs could act as a forum for information sharing, reaching agreement and developing recommendations on how to solve regional air pollution problems. The new regional organization could provide technical support and assessment, and create areas of influence (AOI) and areas of violation (AOV) using air quality modeling and tracer experiments. Institutional mechanisms also could be structured to support the development and implementation of incentive- and market-based approaches to managing regional pollution problems, including developing positive incentives for upwind areas to reduce precursor emissions, such as emissions trading and air pollution funds. The organization also may endow areas of violation with some power to compel actions from areas of influence. Under RAMPs, states would retain primacy, subject to EPA oversight and Federal Implementation Plan (FIP) authority, to the greatest extent consistent with air quality and equity goals, with responsibility assigned at the lowest level of government practicable.

Assuming that the preferred approach to attaining the new NAAQS involves establishing regional emissions cap-and-trading programs, RAMPs could oversee the orderly transfer of emissions credits between jurisdictions, including developing protocols for tracking, verifying, recording, and otherwise overseeing the conditions of interstate and other interjurisdictional emissions reduction credit transactions. RAMPs also could be responsible for reviewing and approving each state's regulations into its SIP for attaining the NAAQS for ground-level ozone.

For this idea to work with the current regulatory system, state SIPs would need to incorporate policies consistent with their RAMPs. EPA could compel states to do this by rejecting the SIPs of all states in the airshed unless they had satisfactory measures enforcing this incorporation. Unless RAMPs have some authority, this arrangement is unlikely to be efficient, because there is no guarantee that a state will agree to bear more costs than other states.

Federal Responsibility with a Diminished SIP Process

A much more radical option would be to dismantle much of the SIP process, except for assuring no net increase in local emissions, and transfer authority for meeting standards to the federal government. This option has several advantages over the first approach. First, federal measures for reducing emissions, such as regional trading options and fuel quality regulations, are almost universally accepted as both the most effective and the most cost-effective. Conversely, state and local measures, such as transportation control measures, have performed poorly. Second, the alternative of establishing RAMPs is problematic, because our federal system does not favor conferring authority to regional, as opposed to state, institutions.

> *A much more radical option would be to dismantle much of the State Implementation Plan process, except for assuring no net increase in local emissions, and transfer authority for meeting standards to the federal government.*

Third, eliminating much of the SIP process reduces costs and contentiousness. This overly bureaucratic and legalistic process draws attention and resources away from the more germane issue of ensuring progress toward the goal of meeting the NAAQS. Furthermore, the enormous modeling uncertainties in demonstrating attainment of the ambient standards could be eliminated and transferred instead to the federal government, which has the expertise, money, and economies of scale to do a better job. And given the large role for regional transport, the federal or regional level is the appropriate one for modeling and demonstration of attainment.

Fourth, forcing localities to demonstrate that their emissions are not increasing is a much more tractable and measurable task than requiring them to demonstrate attainment. Localities likely would still keep their inspection and maintenance programs in place, for instance, and could be required to show that the program was being well implemented. Fifth, the public's attention could turn to the government body that actually has the power to address the problem—the federal government—rather than the localities, which are largely helpless to further reduce concentrations.

The main drawback to this approach stems from its radical departure from the CAA. Congress probably would need to act. Also, many stakeholder groups would be skeptical of such a change at first, with perhaps the greatest concerns being about delay, backsliding on local emissions reductions, and the process by which transportation planning and air quality planning are required to be coordinated. This is why the nondegradation requirement would be so important.

A.J.K.
J-S.S.

Redirecting Superfund Dollars

by Katherine N. Probst

In 2004, the U.S. Environmental Protection Agency's (EPA) appropriation for the Superfund program was almost $1.3 billion dollars. Over the course of your administration, we can expect government to spend an additional $5.2 billion. Mr. President, for the integrity of the program and for the public interest, I recommend that your administration conduct a thorough evaluation of the current allocation of resources within the Superfund program. Based on that evaluation, your administration can recommend which expenditures are truly necessary for the successful implementation of the program and determine where a redeployment of resources—in both staff and dollars—might better ensure that we accomplish the central goal of Superfund—that is, thorough and expeditious cleanups of sites across the nation that are contaminated with hazardous substances. This "zero-based budgeting" approach must be conducted by people who are knowledgeable about the various components of the Superfund program but whose budgets will not be affected by the outcome of the evaluation. It is critical that the evaluation not become an excuse for dismantling the program but truly be aimed at how to best deploy limited resources to achieve the goals of the program—cleaning up contaminated sites and protecting public health.

Background

In 1980, when Congress first enacted the Comprehensive Environmental Response, Compensation, and Liability Act (CERCLA)—better known as Superfund—few could envision the current program. Today, EPA's National Priorities List (NPL) includes more than 1,200 sites. The average cost of cleaning up an NPL site is approximately $25 million, with almost 10 percent of the sites costing $50 million or more to clean up, and some costing hundreds of millions of dollars.

CERCLA created a two-pronged approach to getting contaminated sites cleaned up. Under the statute, EPA has the power—and the money—to conduct site cleanups itself. Superfund also created a far-reaching liability scheme whereby

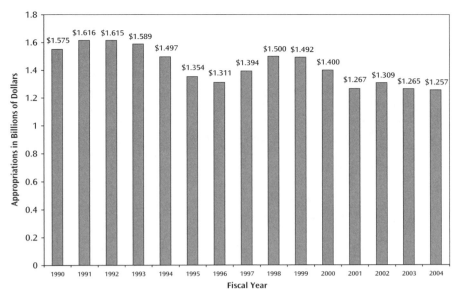

Source: U.S. EPA.

Figure 14-1

Superfund Appropriations by Fiscal Year

those responsible for contamination, referred to as "responsible parties," are subject to retroactive, strict, and joint and several liability. The liability scheme that many love to hate has, in fact, been quite successful in promoting cleanup. Seventy percent of cleanups are implemented directly by responsible parties as a result of EPA's enforcement program. The remaining cleanups are paid for by EPA with funds included in the agency's annual appropriations from Congress.

Since 1990, congressional appropriations to EPA for the Superfund program have ranged from a low of just under $1.3 billion (current levels) to a high of $1.6 billion in fiscal years 1991 and 1992, as shown in Figure 14-1.

In FY2002, for the first time, it became clear that EPA faced a significant funding shortfall and did not have the moneys to pay for cleanups that were ready to go. In FY2004, EPA still faced a major shortfall of funds for cleanups. According to a February 2004 report by EPA's Office of Inspector General, the cleanup backlog then totaled $178 million. Adding support to the argument of a funding shortfall is the fact that the president's budget request to Congress for both FY2004 and FY2005 included an additional $150 million in funding expressly for cleanups. Congress provided only a small portion of these funds in the FY2004 appropriation.

As a result of this funding shortfall, some communities with contaminated sites have waited for years, in some cases for decades, only to find that the cleanups they had been waiting for have been put on hold, perhaps indefinitely. This threatens the credibility of the nation's most visible cleanup program. In addition, the longer it takes to clean up sites, the greater the cost to EPA, as sites must be monitored while awaiting further funds.

To reduce this funding gap, EPA needs to carefully scrutinize how Superfund resources are spent to make sure that the available funds are used in the most effective way to assure high-quality and expeditious cleanups. The agency needs to examine the current deployment of resources, including staff and other funds, and evaluate which currently funded activities are most critical to the program's mission of cleaning up sites and reducing threats to the environment and public health. Whoever is responsible for this task must understand the complexities of the Superfund program so that he or she will make wise decisions. For example, it would be foolish to cut enforcement resources, as money spent on enforcement results in cleanups paid for by responsible parties and ultimately saves the federal government money.

A number of difficult questions need to be asked: Are all the enforcement resources actually going to enforcement activities? Has EPA's technology innovation program provided enough benefits for the agency's investment? Are all the staff paid for by the Superfund program actually working on Superfund activities, or are they being used for other activities and programs? Answering these types of questions will take attention to detail, precision, and guts.

That said, it is almost certain that EPA will not be able to reallocate enough staff and funds to address a $200 million, and most likely growing, shortfall. Thus the major question facing Congress regarding the Superfund program may well be whether Superfund appropriations should be increased. However, it is unlikely that there would be broad-based support for increased congressional appropriations, unless or until EPA gets its own house in order and shows that it has done the best job it can to redirect program funds to make sure that current Superfund resources are being deployed in the most effective fashion. In sum, it is critical to assess where Superfund dollars are going, and whether it is possible to direct a larger percentage of these funds to actual cleanup without crippling other key aspects of the program.

> *It is critical to assess where Superfund dollars are going, and whether it is possible to direct a larger percentage of these funds to actual cleanup without crippling other key aspects of the program.*

Prior Efforts

A congressionally mandated study that was issued by Resources for the Future (RFF) in 2001 titled *Superfund's Future: What Will It Cost?* found that current funding levels would not be adequate for successful implementation of the program under current policies. Estimates of total funding needs for the ten years from FY2000 through FY2009 ranged from just under $14 billion to $16.4 billion. Although the RFF report included year-by-year estimates of funding needs, the congressional mandate did not include an assessment of how resources should be reallocated within the program.

In recent years, a number of other independent reports have been issued by the Congressional Budget Office, the Congressional Research Service, the U.S. General Accounting Office, and EPA's Office of Inspector General regarding how Superfund resources are spent, funding shortfalls for cleanups, and future funding needs. All point to the need to better understand how that portion of Superfund expenditures that goes to non-site-specific activities—about 50 percent of the $1.3 billion—is spent, and whether some of these funds could be redeployed to site cleanup activities.

Although several of these reports have recommended that EPA examine the details of program spending to identify specific opportunities for reprogramming funds to increase spending on site cleanup and other critical program activities, no one has yet conducted this kind of detailed evaluation. Conducting this kind of fundamental assessment of spending patterns and priorities is extremely difficult within the federal government, where, typically, the reallocation of resources is done in an incremental fashion. In May 2004, EPA released the results of an internal study, *Superfund: Building on the Past, Looking to the Future*, requested by the agency's deputy administrator. The express purpose of this internal evaluation was to "identify program efficiencies that would enable the Agency to begin and ultimately complete more long term cleanups … with current resources." The report includes more than 100 recommendations and a lot of data, but the authors fail to make any specific suggestions about what current activities could be cut or reduced, a necessity if spending for cleanup is to be increased.

The next effort to deal with the details will be an audit requested by Congress in the conference report to the FY2004 EPA appropriations. Congress directed EPA's Office of the Inspector General to evaluate Superfund expenditures and to recommend how to increase dollars going to cleanup and minimize administrative costs. The EPA Office of the Inspector General expects to issue two reports in early 2005. The first is expected to highlight areas where administrative efficiencies could be achieved. The second would include broader recommendations aimed at increasing program effectiveness. It is unclear whether even the Inspector General's office, which received $13 million in annual appropriations in the FY2004 Superfund budget, will have the fortitude to identify specific areas where expenditures could—or should—be decreased. It may well be that the needed hard-hitting analyses will be conducted and the truly difficult recommendations made only if a similar effort is directed by an organization outside EPA that answers directly to the White House.

The Superfund program is important to the American people. It is critical that your administration take the steps needed to assure that the program does the best possible job in accomplishing its goals of cleaning up contaminated sites and protecting public health and the environment.

Although a number of reports have highlighted the future funding needs of the Superfund program and made suggestions for management and efficiency reforms, none have addressed the more basic questions of how Superfund dollars are being spent and whether fundamental (or marginal) changes should be made in how resources—both intramural and extramural—are allocated. This task has taken on more urgency now that sites across the country face demonstrable and visible delays in cleanups.

The Superfund program is important to the American people. It is critical that your administration take the steps needed to assure that the program does the best possible job in accomplishing its goals of cleaning up contaminated sites and protecting public health and the environment. The zero-based budgeting approach proposed here is an important step in assuring that it does so.

K.N.P.

15

A Broader View of Brownfield Revitalization

by Kris Wernstedt

S ir, a policy change is needed to refocus federal resources to larger-scale, areawide revitalization of brownfield sites—underused properties whose redevelopment is complicated by the fact or perception that contamination is present. The cleanup and redevelopment of such land can both revitalize neighborhoods burdened with a legacy of contamination and enhance tax revenues. However, most cleanup and redevelopment to date has rested on a narrow, property-by-property approach. Your initiative should include a limited-term amnesty period to encourage the volunteering of contaminated properties for a large-scale revitalization effort, strong public participation and reporting requirements, innovative financing mechanisms, and an expansion of the range of sites eligible for assistance.

Background

A variety of federal and state laws apply to brownfield sites. Fearing cleanup liability under these laws, most notably the federal Superfund statute, many owners and prospective buyers of brownfields remain leery of redeveloping or even investigating such properties and attracting regulatory attention. The parties who may have caused the contamination may be long gone or unable to pay for cleanup, and few of the brownfield sites would qualify for public funding under the Superfund program or its state counterparts. Yet left unattended, brownfield sites may pose threats to public health and the environment and depress the economy of local neighborhoods. Moreover, existing roads, utilities, and other public infrastructure may remain underutilized even as undeveloped rural or suburban properties attract development, contributing to social and economic segregation and continued underinvestment in older urban and industrial areas. This problem bedevils communities across the country, with hundreds of thousands of acres at stake. The U.S. General Accounting Office reports that the nation has between 130,000 and 450,000 brownfield sites, based on data collected from several federal agencies in the 1980s. Other estimates put this number as high as 1 million.

Throughout the 1990s, federal and state legislative and regulatory reforms reduced some of the barriers to redeveloping brownfield sites, and at the programmatic level, the Brownfields Program of the U.S. Environmental Protection Agency (EPA) supported hundreds of pilot projects that resulted in the environmental assessment of thousands of properties. Moreover, the signing of the Small Business Liability Relief and Brownfields Revitalization Act in early 2002 has provided firmer statutory footing for expanded liability protection and additional federal funding. In 2003, EPA awarded $75 million in grants to states, local governments, and nonprofits under the new law, bringing its investment in brownfields since 1995 to more than $700 million. In addition, all but a handful of states now offer voluntary cleanup programs to encourage owners and developers to come forward and voluntarily address site contamination, in exchange for less onerous requirements and certainty that state authorities will not continue to hold them liable for additional cleanup.

An Alternative Approach

Federal, state, and local brownfield redevelopment efforts generally aim at supporting the highest and best use of individual properties. Although these efforts have in many cases yielded significant increases in jobs and tax revenues on a property-by-property basis, they have not been as successful at revitalizing whole communities. The piecemeal approach is particularly problematic in areas with a large number of scattered properties that are less than a half acre in size. These properties often attract little interest from the private sector because of their location in distressed areas, the high fixed costs of remediation and redevelopment relative to the small gross returns on such small properties, and the lack of cost-effective insurance options to reduce the risk of encountering unexpectedly high costs.

An alternative areawide approach that explicitly treats multiple brownfield properties as a system and tackles them en masse rather than each in isolation offers a number of potential benefits and could improve the prospects for community revitalization in four related ways. First, redeveloping multiple small brownfield properties in a coordinated fashion can be financially attractive to both public and private developers, particularly for residential reuse. For the public, the cumulative effects of redeveloping multiple properties can increase property values, tax revenues, and other community benefits over an entire neighborhood depressed by a small number of contaminated sites. Additionally, increases in property values over a neighborhood may increase the expected market price for new housing to a high enough level that a private developer will undertake a new project. Even absent sufficiently high market rates, anticipated public benefits might justify subsidies that reduce investment risks and provide the developer with an acceptable rate of return.

Redeveloping multiple small brownfield properties in a coordinated fashion can be financially attractive to both public and private developers, particularly for residential reuse.

Second, an areawide approach may allow larger dollar investments in cleanup and redevelopment that can take advantage of both economies of scale in remediation and infrastructure for redevelopment and risk-sharing opportunities across multiple sites. For example, the cost of investigating contamination and remediating 10 properties in a coordinated fashion may be far less than doing each in isolation, particularly if the properties share a common environmental problem, such as underlying groundwater contamination, and would benefit from a similar remedial strategy. Even if the contamination problems are dissimilar, the exceptional risk of one property encountering unanticipated and potentially costly cleanup surprises may be balanced by the unexceptional risks of the other nine properties.

> *The cost of investigating contamination and remediating 10 properties in a coordinated fashion may be far less than doing each in isolation, particularly if the properties share a common environmental problem and would benefit from a similar remedial strategy*

Third, if multiple contaminated properties within a defined area can be put under single ownership, an areawide approach may make environmental insurance a financially critical part of a redevelopment strategy. Such insurance can provide protection against unanticipated overruns in cleanup costs, as well as liability protection for a wide range of risks, such as the discovery of additional contamination or lawsuits by site workers or adjacent property owners for personal injuries suffered as a result of the contamination. The insurance is currently unaffordable for small projects, but bundling multiple properties together could make it cost-effective.

Finally, an areawide approach can place the redevelopment of brownfields in a comprehensive, integrated planning framework that takes advantage of the opportunities these contaminated properties provide. Rather than limit the benefits of redevelopment to the sum of the benefits from individual projects, a community undertaking an areawide approach can design the redevelopments so that they are complementary and synergistic, causing the value of the whole to exceed the sum of the parts.

Options for Areawide Approaches

EPA has recognized the promise of areawide approaches in two related initiatives it announced in 2003. In the Land Revitalization Agenda, the agency identified areawide assessments as one approach for reducing costs and called for the coordination of grants to support the redevelopment of clusters of contaminated properties. More broadly, in both its Land Revitalization Agenda and the One Cleanup Program, EPA has provided modest funds to support areawide pilot projects that promote coordinated approaches for cleaning up areawide contamination problems across its multiple land cleanup programs.

Several states recently have ventured into areawide strategies. For example, a 2002 policy directive from New Jersey's Department of Environmental Protection called for the establishment of an areawide brownfield development program to enable communities in the state to plan comprehensively for the remediation and reuse of multiple sites. To support this, the department has developed a "cluster approach" to address sites located near each other. In neighboring New York, a state law signed in 2003 includes provisions to create and fund Brownfields Opportunity Areas. This program encourages community groups to undertake areawide planning and provides incentives for private developers to participate in this effort.

States and nonprofit conservation entities also have explicitly supported brownfield projects that promote community and environmental benefits beyond the confines of an individual contaminated and underutilized parcel. Such projects deemphasize the job and tax benefits that traditional brownfield economic development programs promote in favor of broader gains. For example, the redevelopment of a contaminated parcel and its conversion to open space or parkland may provide environmental benefits such as improved water quality, by virtue of limiting the runoff of polluted water, and increase the value of surrounding properties. Wisconsin's Brownfields Green Space and Public Facilities Grant Program exemplifies this broader appeal with a stated intent to assist communities with the financial costs of the cleanup of brownfield properties that will be redeveloped into community assets that yield public benefits.

These various federal and state areawide approaches have brought some important new aspects of areawide redevelopment to the policy world, but none of them have fully embraced the potential of this approach. The EPA actions are pilot efforts and arguably aim not so much at areawide revitalization as at the integration of EPA's major cleanup programs across multiple sites or broad areas of contamination. The few state programs with areawide programs also have not focused on revitalization per se, in part because they typically rely on tactical targets of opportunity—larger properties with significant development potential—that will yield benefits in a relatively short time.

A strategic program of investments in smaller properties with a longer-term planning horizon for the community has proven to be a difficult political sell in an environment of scarce public funds. Similarly, it has been difficult to garner widespread support for open-space reuses of contaminated properties, a use that has great potential for areawide revitalization, largely because of an understandable desire to realize more immediate gains in jobs and on-site property tax revenues from redevelopment projects.

Recommendations

I recommend that you develop a federal areawide revitalization initiative that specifically targets the redevelopment of multiple smaller brownfields in distressed urban and rural communities. Specifically, the areawide initiative should include the following six substantive elements:

■ new criteria in EPA brownfield grants that reward an applicant who presents a redevelopment proposal that includes multiple small parcels of contaminated land and documents the expected spillover potential from each parcel for providing communitywide environmental, health, and social benefits;
■ strengthened requirements for public participation in areawide projects that receive federal support, including funding to promote the involvement of community groups;
■ a limited-term amnesty that provides partial liability waivers, grants for site assessments, and matching grants, loans, or insurance subsidies for site cleanup

to encourage owners of small, "mothballed" contaminated parcels to perform site investigations as a prelude to redeveloping or marketing the properties;

■ changes in the federal brownfield tax incentive, which lets taxpayers who incur cleanup expenses fully deduct these costs from their taxable income in the year they are incurred, to make the incentive permanent, allow petroleum sites to take advantage of the incentive, and eliminate the requirement that the expenses be recaptured when the property is sold;

■ provisions in the tax incentive that will allow trading of the tax deduction for cleanup expenses, so that qualified brownfield projects for which the incentive currently is not attractive (such as sites with activities that do not produce income or those owned by local governments) can benefit by selling their allowable deduction to other parties; and

■ mandated requirements that all properties receiving federal support report data on contamination and assessment and remediation costs for use in the development of a publicly accessible database to provide better information on the environmental risks of brownfield redevelopment.

These recommended changes would correct the myopia of piecemeal, site-by-site brownfield redevelopment, both through direct efforts that specifically support an areawide perspective and by indirect means that increase the range of contaminated sites attractive for redevelopment and the range of parties interested and able to participate. For example, allowing the sale of tax credits could open up the possibility for local nonprofit, municipally affiliated entities to actively pursue the remediation of tax-delinquent contaminated properties, financed in part by the sale of the tax credits.

> *Allowing the sale of tax credits could open up the possibility for local nonprofit, municipally affiliated entities to actively pursue the remediation of tax-delinquent contaminated properties.*

The proposed initiative will require a financial commitment to areawide brownfield revitalization through grants, loans, and other financial incentives. If the federal budget deficit and offset requirements militate against an increase in federal support, I recommend a middle political meeting ground of refocusing existing allocations to promote the six substantive elements of the areawide brownfield initiative I have described.

K.W.

Modernizing the Food Safety System

by Michael R. Taylor

Fundamental reform of the nation's essential yet tradition-bound food safety program requires presidential leadership. Mr. President, I recommend that you organize a bipartisan effort to design the food safety system of the future and implement it in a way that builds on current strengths but prepares the system for future success. While the U.S. food safety system has many strengths, over the century plus of its existence it has become organizationally fragmented, bound by obsolete statutes, and unable to make the best use of its scarce resources to protect the safety and security of the American food supply. To meet the persistent challenge of foodborne illness and the new challenges posed by the globalizing food system—illustrated by increased dependence on imports, the emergence of mad cow disease, and the threat of bioterrorism—fundamental reform is required. This reform should do several things: focus the system more effectively on prevention of foodborne illness; improve accountability across the system for meeting science-based food safety performance standards; establish an integrated food safety strategy and leadership structure; and emphasize risk-based resource allocation.

The Challenges

Calls to modernize the food safety system have come in numerous reports issued by the National Academy of Sciences (NAS) and the General Accounting Office, which point out that more than a dozen federal agencies have major responsibilities for food safety, but that the system lacks a focal point for food safety leadership, responsibility, and accountability. The U.S. Department of Agriculture (USDA) regulates meat and poultry products and processed eggs. The Food and Drug Administration (FDA) regulates all other foods, including seafood, dairy products, and shell eggs, except that the Environmental Protection Agency decides how much pesticide residue can be present in food, and the National Marine Fisheries Service in the Department of Commerce conducts an extensive fee-for-service inspection program for seafood. The Centers for Disease Control and Prevention (CDC) conducts surveillance of foodborne illness, while the FDA, multiple components of USDA, and the Department of Homeland Security all have important food safety research programs.

This fragmentation reflects the unplanned accretion of programs and activities over the years and is grounded in statutes that contribute to, and in some instances compel, inconsistent and ineffective use of resources. For example:

- USDA individually inspects every one of the 7 billion poultry carcasses produced annually in the United States using a mandated approach that the NAS discredited nearly 20 years ago as largely ineffective in detecting real safety hazards, whereas the FDA struggles to inspect "high-risk" seafood plants once a year.
- USDA sends inspectors to every foreign meat plant producing for the U.S. market to verify that products are produced and inspected in accordance with U.S. standards, whereas the FDA depends almost entirely on border inspections that cover only 1 to 2 percent of import shipments and rely on inspectors detecting contamination problems after they have occurred.
- Some food safety problems and prevention opportunities fall entirely through the cracks. USDA attempts to address the dangerous pathogen *E. coli* O157:H7 at the meat slaughter and processing stage, and the FDA is responsible for the bacteria when it turns up in fresh produce, but neither agency has a mandate to tackle the problem on the farm, where *E. coli* O157:H7 originates in the gut of cattle and preventive measures are most likely to be effective.

Foodborne illness remains an important public health problem in the United States, with the CDC estimating 5,000 deaths and 325,000 hospitalizations annually, virtually all of which are preventable through better-targeted research, regulation, and education.

These and many other well-documented inconsistencies and gaps in the food safety system are themselves sufficient grounds for modernizing the system. They are part of the reason why foodborne illness remains an important public health problem in the United States, with the CDC estimating 5,000 deaths and 325,000 hospitalizations annually, virtually all of which are preventable through better-targeted research, regulation, and education.

But the new threats posed by America's place in today's global food system raise the stakes even higher, in both public health and economic terms. Open international markets make wider choices available to American consumers so that we can enjoy fresh produce year-round and have competitively priced options for meeting both nutritional needs and food preferences. These benefits come, however, with dependence on how food is grown, processed, and inspected elsewhere, and with the possibility of harmful contamination being introduced beyond the reach of any direct U.S. oversight. In the winter months, most fresh fruits and vegetables consumed in the United States come from other countries, mostly in Latin America. Year-round, 70 percent of U.S. seafood is imported. To maintain the benefits of a diverse, economical food supply and to protect public health, the U.S. food safety system needs to modernize its approach to ensuring the safety of imported food.

Mad cow disease and bioterrorism further illustrate the new challenges confronting the United States as a key participant in the global food system. A single infected cow in Washington State disrupted trade with Canada and blocked all beef exports to Japan, with significant economic consequences for American agriculture. Bioterrorism looms over the food supply as a threat to public health if not prevented, and as a potentially severe disruptor of the American economy if any incident is not contained and responded to swiftly, coherently, and in a manner that restores public confidence in the safety and security of the food supply.

On both mad cow disease and bioterrorism, the U.S. government responds today through ad hoc coordination among a fragmented patchwork of agencies with conflicting missions concerning food safety, the economic welfare of agriculture, trade policy, and homeland security. When it comes to the safety and security of the food supply, and maintaining confidence among American consumers and trading partners, a single official and agency with both a clearly defined food safety mission and accountability for success is needed.

> *On both mad cow disease and bioterrorism, the U.S. government responds today through ad hoc coordination among a patchwork of agencies with conflicting missions.*

Recommendations

Modernization of the food safety system to meet these challenges requires major statutory and organizational change. In our system of government, however, broad-based reform too often comes only in response to real or perceived crises. That certainly has been so in the case of many U.S. health and environmental programs, and we have seen in the United Kingdom and other countries the political potency of food safety in times of crisis. In response to its mad cow crisis, the United Kingdom scrapped its old multiagency food safety structure and unified its food safety functions in the newly created Food Standards Agency.

In the United States, you as president have an opportunity to read the signals about the shortcomings of the U.S. system and provide leadership for change in advance of any catastrophically disruptive crisis. Recognizing the difficulty of making sweeping reforms through the normal legislative process and the need to unite diverse stakeholders behind a reform agenda, you should exert your leadership through a three-step process involving design, legislation, and implementation of reform.

> *Modernizing the food safety system requires major statutory and organizational change.*

Design

You should establish a Bipartisan Presidential Commission on Food Safety Reform to design statutory and organizational reform in accordance with principles and objectives that you enunciate. The commission should be formed in consultation with congressional leaders of both parties and include a prominent and diverse group of food safety leaders, experts, and stakeholders. Your charge to the commission should be grounded in the expert consensus reflected in the numerous NAS and GAO reports and point toward a food safety system with the following attributes:

■ **Focus on prevention.** Effective prevention of food safety hazards and foodborne illness is the key to protecting public health and maintaining the confidence of consumers and trading partners in the safety of the American food supply. The food safety agencies have embraced the principle of prevention by mandating the adoption of Hazard Analysis and Critical Control Points (HACCP) in meat, poultry, seafood, and juice processing plants, but they have no statutory mandate and uncertain legal authority to build a system that works to prevent the introduction of food safety hazards across the spectrum from the farm to the table.

■ **Clear standards of accountability.** An effective system of prevention depends on the government being able to set food safety performance standards that establish a standard of care that companies can be held accountable for meeting. This principle is well established for chemical hazards, but industry challenges of USDA's initial attempts to apply the concept to microbial hazards have revealed the obsolescence of current food safety laws, which were adopted before microbial pathogens came to the fore as a public health problem of central concern to food safety regulators.

■ **Integration of strategy and leadership.** Experts widely recognize that prevention of foodborne illness and management of such problems as mad cow disease and bioterrorism require an integrated, systems approach from farm to table and should harness the tools of research, regulation, and education in a coherent strategy. This is made impossible by the current organizational fragmentation of the system, which fragments food safety leadership and defeats accountability for the system's successes and failures. Fragmented leadership domestically also undercuts U.S. food safety leadership internationally, which has negative health and economic implications.

■ **Risk-based resource allocation.** In the end, the success of the food safety system depends on making the best possible use of resources to target and minimize the hazards that pose the greatest risk to public health and the overall safety and security of the food supply. The food safety agencies have no congressional mandate for risk-based resource allocation, and the meat and poultry laws, which control the allocation of at least 60 percent of federal food safety resources, largely preclude it.

Legislation

Based on the legislative and organizational design developed by your commission, you should provide leadership for a legislative process to replace the current food safety laws with a modern, unified law. A successful legislative process requires that diverse stakeholders benefit in diverse ways from the outcome. Your leadership role in this process is to help all stakeholders see the mutual advantages of food safety reform while ensuring that the legislative outcome is faithful to the principles and objectives embodied in your charge to the commission. You also should prepare a resource plan for consideration by Congress to ensure that the funding provided for the new system is at an adequate level and comes with the flexibility needed to make the system a success.

Implementation

The transition to a new statutory and organizational structure is an enormous management task. Such change is inherently costly and disruptive. You should devise a five-year transition plan that minimizes the cost and disruption, avoids lapses in protection during the transition, and prepares the system for success.

Cautionary Notes

Virtually universal agreement exists that if we were starting from scratch, we would not design the food safety system the way it is today, with multiple agencies and with statutes that both lack a modern food safety mandate and contain old mandates that block risk-base resource allocation. The principle opposition to change comes from those who argue that the current system is working well enough and that the costs and disruption of transitioning to a new system outweigh the benefits. This argument needs to be taken seriously in the formulation of the implementation and transition plan, but it overlooks the fact that the costs of reform are incurred in the relative near term, whereas the benefits will be enjoyed over the long term. If the functioning of the system over the next five to ten years were the only concern, we might just try to do the best we can with the tools we have and hope that a major food safety crisis does not occur during that period. But if the concern is for the longer term, we should be more willing to invest now in change that will return benefits for years to come.

This reality of the long-term nature of the benefits of food safety reform is why presidential leadership is so important. Some of the strongest resistance to change has come from the food safety agencies themselves, whose managers and leaders are understandably focused on the unrelenting stream of immediate problems that they are responsible for addressing today and every day. It is unrealistic and unfair to expect the agencies to lead the way on fundamental structural reform.

> *It is unrealistic to expect agencies to lead the way on structural reform. It is rather for society, through political leadership, to judge whether the country's long-term health and economic well-being will be served by sticking with the status quo.*

It is rather for society, through its political leadership, to judge whether the current system is performing well enough and whether the country's long-term health and economic well-being will be served by sticking with the status quo on food safety. Among our political leaders, you as president are best situated to make that judgment, to make the case publicly for change, and to lead the process. Without your leadership, it is unlikely that change will occur, at least until circumstances force change in an atmosphere of crisis.

M.R.T.

Performance Standards for Food Safety

by Sandra A. Hoffmann and Alan J. Krupnick

Mr. President, we urge you to adopt broad and consistent use of product performance standards as the centerpiece of foodborne pathogen regulation. Performance standards help assure product safety, maintain consumer confidence, and provide economic incentives for industry to find more efficient means of meeting food safety goals. This action would build on and strengthen already accepted principles of food safety regulation. In some instances, it will require legislative change; in other cases, sufficient legal authority already exists. Our recommendation draws on and is in large part in agreement with recent recommendations from the 2003 National Academy of Sciences Committee on Review of the Use of Scientific Criteria and Performance Standards for Safe Food.[1] Establishing performance standards is one element of the overall structural reform recommended in Chapter 16. Even if the full menu of reforms suggested there cannot be implemented, it should be possible to implement clear performance standards.

Background

Since at least the early 1980s, major advisory bodies, including the U.S. General Accounting Office (GAO) and the National Academy of Sciences (NAS), have made repeated calls for establishing a clear relationship between public health outcomes and food safety regulatory requirements. Much of the U.S. statutory food safety inspection mandate dates from the early 1900s and now serves very little public health purpose. For instance, the visual inspection of animals and carcasses required by the meat and poultry inspection acts are incapable of detecting or preventing pathogen contamination that leads to thousands of deaths and hundreds of thousands of hospitalizations each year in the United States.

These same advisory bodies have insisted on the need for greater flexibility to allow food processors to take advantage of rapid advances in biological sciences and food process technology. This insistence is a reaction to use of a command-

and-control approach to food safety regulation, such as mandating a particular food safety technology or manufacturing process. In the 1990s, the U.S. Department of Agriculture (USDA) and the Food and Drug Administration (FDA) moved, within the limits of statutory and political constraints, to implement a more modern process standard approach to assuring product safety. They required meat, poultry, seafood, and juice processing plants to develop and implement a risk-management process: agency-approved Hazards Analysis and Critical Control Point (HACCP) plans. In developing these plans, plant management must access their production process, identify critical points where hazards could enter the system, and develop and implement a plan to control those critical points. However, most HACCP plans do not say what the plants should be trying to achieve through this process. The use of such plans was intended to encourage manufacturing processes that prevent product contamination with pathogens and limit pathogen growth. The major goal of these reforms was to facilitate adoption of modern quality-control processes. An equally important goal was to allow industry the flexibility to choose how to meet these goals.

The visual inspection of animals and carcasses required by the meat and poultry inspection acts are incapable of detecting or preventing pathogen contamination that leads to thousands of deaths and hundreds of thousands of hospitalizations each year in the United States.

But use of HACCP plans is falling short from the perspective of providing the flexibility needed to encourage efficiency and innovation. A plant's plan must be preapproved by the regulatory agency before it can be implemented. The structure of requiring government preapproval of production processes at each plant limits a plant manager's ability to take advantage of new technology and the learning that occurs in implementing a production process. Although plans are to be revised on a periodic basis, preapproval inherently imposes constraints on firms' flexibility to find better ways to control food safety hazards. Furthermore, experience with environmental regulation suggests that plants will tend to follow agency guidance regarding choice of process to reduce their regulatory risk. This behavior significantly reduces the actual flexibility provided by a process standard approach like HACCP. In addition, both the FDA and Food Safety Inspection Service (FSIS) are statutorily required to regulate at the plant level. This plant-by-plant approach makes it impossible to cost-effectively manage risk along the entire supply chain. For example, it may be more cost-effective to minimize salmonella in poultry flocks than it is to decontaminate poultry carcasses.

Furthermore, current HAACP regulations still are not tied to desired public health outcomes. Plants are required to implement plans that reduce pathogens to an "acceptable level." But the regulations do not define what an acceptable level is or how it is related to public health outcomes. A recent NAS committee noted that current regulations give "little or no guidance on the level of hazard control expected by the government from an adequately designed and implemented HACCP plan." The GAO and the interested public have criticized HACCP regulation as providing inadequate accountability for food safety. Plants can be in compliance with agency-approved process control plans, but because the plans are not tied to achieving expected health outcomes, no way exists to assure that they will meet public health goals.

Plants are required to implement plans that reduce pathogens to an "acceptable level." But the regulations do not define what an acceptable level is or how it is related to public health outcomes.

Tying HACCP regulation to product performance standards based on public health impact could address both of these shortcomings. In 2003, an NAS committee recommended requiring food safety product performance standards tied to explicit public health goals as a way of both assuring that food safety regulation achieves its public health purpose and providing incentives for innovation and efficiency.

The Product Performance Standard Approach

A system of performance standards should include several elements.

Accountability Tied to Public Health Goals

Food safety product performance standards need to be tied directly to accomplishment of a clear public health goal. The public health goal could be a desired reduction in incidence of foodborne disease or an acceptable risk of illness. EPA uses such a goal—not exceeding a one-in-a-million risk of cancer—in pesticide registration. The public health goal should be translated into measurable food safety objectives (FSO). The 2003 NAS report defines an FSO as "the maximum frequency and/or hazard in a food at the time of consumption that is considered tolerable for consumer protection." By using data, scientific modeling, and where necessary, professional judgment, it is possible to determine the concentration of pathogen in product (the performance standard) at any given point in the supply chain that can be reasonably expected to result in a contamination level at the time of consumption that meets the FSO.

Incentives for Efficiency and Innovation

The decision of where in the food supply chain to impose product performance standards will affect the incentive they provide for efficiency and innovation. In theory, the greatest efficiency gains can be achieved by imposing a single performance standard as close to final consumption as possible. For example, an *E. coli* O157:H7 standard could be imposed on ground beef as it enters the retail chain. This would allow market forces to create incentives for wholesalers, shippers, processors, and producers to take control efforts at the most cost-efficient points in the whole production–processing–marketing chain. It also would create incentives for firms to innovate and create processes, procedures, and technologies that reduce costs.

In theory, the greatest efficiency gains can be achieved by imposing a single performance standard as close to final consumption as possible.

Final suppliers, such as McDonald's, already can demand specific safety levels from its meat suppliers. This and other experiences with "channel captains" in the food industry demonstrate that it is possible for product standards placed at the end of the supply chain to provide incentives for cost-effective reduction in pathogen risk throughout the supply chain.[2]

Given current statutory constraints, NAS suggests establishing a series of product performance standards for transportation and retail, processing, and on-farm.

This is less advisable than a single standard as far forward as possible in the chain, perhaps one at the processing plant and another at delivery to retail outlets. If stage-specific performance standards were adopted without consideration of their impact on the entire food supply-and-demand chain, the criteria selected may be inefficient and performance standard–based regulation could create real disincentives for firms to adopt effective food safety controls.

Market forces will tend to promote cost-saving technological change. However, it is less clear that there is as strong a market incentive for safety-enhancing technological change. Performance standards can enhance such incentives if tied to public health goals and FSOs that require continuous improvement.

Science-Based Enforcement

A product performance standard approach to food safety regulation must include a science-based approach to enforcement. Compliance with performance standards can be scientifically verified through either product sampling or established statistical process controls. Modern industrial practices use advanced statistical methods for assuring that a production process meets a given standard for its products. In addition, the regulatory system should allow for adoption of advances in process control verification.

An Ability to Adjust to Rapid Change

Biological sciences, food consumption patterns, and food production practices in the United States all are changing very rapidly. In the past 20 years, we have experienced the impact of emerging pathogens, such as the deadly *E. coli* O157:H7, and have learned that acute infections from foodborne pathogens can have serious long-term health effects. In response, a food safety regulatory system organized around product performance standards must be continually updatable. Public health goals, FSOs, and the performance standards derived from them need to be reviewed on a regular basis without need for regulatory change. However, experience with chronic delays in the U.S. EPA's program of periodic reviews of the scientific support for pesticide registrations warrants examination in designing such a system.

Congressional Support for Data Collection

In order to set public health objectives and to know whether the regulatory system is achieving the goal of protecting public health, good surveillance data that are reliably collected over time are needed. Data also are lacking to conduct some of the risk assessments needed to translate public health goals into FSOs and performance standards. Congress needs to provide adequate resources to develop and maintain data support necessary to design, monitor, and enforce performance standards.

Obstacles

Several obstacles exist to the adoption of performance standards. These include legal, market, technical, and political obstacles, as well as a lack of enforcement resources.

Legal Obstacles

Statutory authority to promulgate product performance standards may be lacking. The attempt by FSIS to use mandatory salmonella product performance standards as a criteria for judging the adequacy of meat processing plants' HACCP plans was struck down by the 5th Circuit Court of Appeals in 2001 as being outside the agency's statutory authority. Bills were introduced in the 108th Congress to give the USDA authority to use mandatory performance standards, but they were not referred out of committee (S. 2013, H.R. 3956).

Placing product standards at the end of the supply chain could require substantial legislative reform. Current federal law compartmentalizes responsibility for food safety both to particular parts of the supply chain and to particular foods. For example, FSIS has jurisdiction over meat and poultry slaughter and processing plants. A frozen pizza could be regulated by several federal agencies. State government has traditionally had jurisdiction over inspections and testing of products in warehouses and at retail. Local governments also have jurisdiction over retail establishments. Creation of a food safety regulatory system built around product performance standards placed on final products could necessitate revisiting these relationships, although it may be possible to introduce performance standards through FDA's Model Food Code, which is revised regularly and has been adopted by all but one state. Use of product performance standards at the point where the product leaves the processing stage and enters the marketing stage of the food supply chain would require much less extensive legislative reform. This portion of the food supply chain currently is under federal jurisdiction.

Market Obstacles

The size of firms in the food industry varies greatly. The design of product performance standards needs to consider the influence of industry concentration on the transmission of signals about product quality requirements upstream from the retailer to the shipper, the processor, and the farm. What works in an industry dominated by a few large retailers, such as McDonald's, may not work in one without large retailers.

As in other regulatory areas, attention needs to be paid to impacts on small businesses, even though they make up a small fraction of total sector output. Small businesses could be given longer deadlines for compliance, exemptions, or more lenient requirements. Information-sharing initiatives also could help raise management and technical sophistication at small firms.

Another central concern is the impact on the food budgets of the poor. The prices of lower-quality products may rise more than the cost of others, as these products now tend to be produced to lower levels of safety. In this case, the poor may no longer have the option of trading off safety for cost. However, the efficiency gains and incentives for innovation provided by performance standards may well reduce prices, putting the poor in a position of getting more safety for a lower price.

Technical Obstacles

Performance standards can work only if they are effectively enforced. Monitoring and measuring contamination are challenging tasks. Currently, we lack data on baseline contamination and efficacy of different interventions. But a performance system creates incentives to improve data. The retailer, as well as the government, has particular incentives to produce data. Indeed, if the enforcement system is designed appropriately, each stage in the farm-to-table chain has incentives to make sure either that the food they pass on is clean or that it is very clear who has the responsibility to clean it.

Despite rapid advances in testing technology, meaningful delays still exist between the time a sample of a product is taken in a plant and the time results are available. By then, the product often has been packaged and even marketed. As a result, as currently conducted, pathogen tests provide a "scorecard" for plant performance rather than a direct indicator of product safety. Faster testing is required.

Potential Political Obstacles

Current inspection laws mandate highly labor-intensive inspection regimes. A major obstacle to past efforts to make major reforms of meat inspection statutes has been resistance from meat and poultry inspectors' unions. Consumer opponents of HAACP view it as "deregulation." Some may have the same reaction to reliance on performance standards to provide incentives to producers. But other consumer groups see product performance standards as a means of assuring accountability for product quality.

Regulated firms may prefer the security of knowing that they are complying with a process standard to the responsibility of having to assure safe product. As experience with environmental regulation suggests, firms will have to be convinced that the agency has conferred on them real compliance flexibility before they will chance a departure from traditional approaches to food protection.

Firms will have to be convinced that the agency has conferred on them real compliance flexibility before they will chance a departure from traditional approaches to food protection.

Lack of Enforcement Resources

Historically, the distribution of resources for enforcement across federal agencies has been uneven. Of particular concern, FDA has had a very low level of funding for inspection activities. The greatest gains in security, efficiency, and innovation from use of food safety performance standards would result from their comprehensive application across the food supply. This kind of new approach to food safety regulation would require reexamination of the distribution of enforcement funding across the federal food safety system.

Conclusions

A shift from reliance on process standards to reliance on product pathogen performance standards could enhance the effectiveness and efficiency of federal food safety regulation. The greatest gains in enhancing the effectiveness and efficiency of the U.S. food safety system would come from creation of performance standards for final consumer products. Such a shift will require legislative action and a major change in the structure of food safety regulation. Smaller, but still significant, gains could be had by using performance standards to regulate pathogens in products as they leave processing plants. This also would require legislative authorization, but it would necessitate a much more modest change in the food safety regulatory structure.

> *A shift from reliance on process standards to reliance on product pathogen performance standards could enhance the effectiveness and efficiency of federal food safety regulation.*

It is critical that product performance standards be based on measurable public health goals, as recommended by the NAS. Without a direct tie to public health outcomes that can be revised and updated in response to changing scientific knowledge, we are in danger of re-creating the same rigidities handed to us by the 1906 Congress.

The following principles[3] outlined by USDA economist Elise Golan encapsulate our recommendations for the design of food safety product performance standards: Regulate as close to the end user as possible to encourage upstream innovation. Choose strict, not merely feasible, standards to encourage efficiency and innovation. And use criteria for compliance verification that are informative, reliably measurable, and flexible.

S.A.H.
A.J.K.

Notes

1. NAS (National Academy of Sciences). 2003. *Scientific Criteria to Ensure Safe Food.* Washington, D.C.: National Academy Press.

2. Golan, Elise, Jean Buzby, Stephen Crutchfield, Paul Frenzen, Fred Kuchler, Katherine Ralston, and Tanya Roberts. 2005. The Economic Value of Foodborne Risk Reduction. In *Toward Safer Food: Perspectives on Risk and Priority Setting,* edited by Sandra Hoffmann and Michael Taylor, Washington, D.C.: Resources for the Future.

3. Golan, Elise. 2004. Performance or Process Standards? What's Best for Food Safety, Efficiency, Innovation, and Trade? Paper presented at Annual Meetings of the American Association for the Advancement of Science. February 2004, Seattle, WA.

Part III
Natural
Resources

Streamlining Forest Service Planning

by Roger A. Sedjo

The Forest Service has a complex and very costly process to create plans for the various national forests. It has been estimated that 40 percent of the total direct work of the agency is involved in planning, which consumes more than 20 percent of the national appropriations, several hundred million dollars. These plans often are not implemented, however, because of appeals and litigation, changing circumstances such as the creation of new critical habitat areas and roadless set-asides, or major disruptions as caused by wildfire. Sir, I recommend that you dramatically revise and simplify the planning process. You can do this without changing the existing legislation, although it may require a revised set of regulations. Simplifying and rationalizing the current forest planning process, which requires a huge amount of detail and the development of very long-term projections, could make planning far less costly, perhaps saving up to 40 percent of the current time and funding requirements. The amount of detail currently included in the plans is unnecessary, as the rapidly changing circumstances of the forest soon render most of the long-term plans obsolete. I propose a scaled-down planning process that would use an Environmental Management Systems (EMS) approach.

Background

Today the National Forest System (NFS), comprising the national forest lands managed by the Forest Service, covers an area of some 192 million acres, mostly in the West. Administratively, it is broken down into 10 regions containing some 150 individual national forests. The Forest Service has an annual budget in excess of $3 billion. In 1988, the NFS produced more than 12 billion board feet of timber; by 2002, harvest had declined to less than 2 billion board feet, about 16 percent of the production of a decade and a half earlier. The national forests currently account for less than 5 percent of U.S. industrial wood production. However, there is not a serious wood shortage in private forests, which are intensively managed and highly productive and now provide more than 90 percent of the industrial wood in the United

States. The remainder of U.S. production comes from other public lands, including local, state, and other federal lands.

The U.S. forest system was established in the final years of the nineteenth century, with the 1891 withdrawal of certain federal lands from the public domain for designation as permanent "forest reserves." This act simply allowed the forest lands to be set aside, with no protective authority or management. A few years later, in 1897, Congress passed the Organic Act, which provided for protection and management authority for the forest reserves. This act mandated the permanent "forest reserve" to do two tasks: secure favorable conditions of water flows, and furnish a continuous supply of timber for the use and necessities of the people of the United States.

In 1905, the Forest Service was created to manage these reserves consistent with the Organic Act. The forest reserves were then called the National Forest System. During the early years of the Forest Service, roughly the period from 1905 until post–World War II, the role of the NFS consisted of relatively simple custodial management: keep poachers out; occasionally harvest some timber, particularly if it would support a local mill; minimize disturbances of watersheds; and make sure that forest regeneration eventually would occur after a harvest, although this was commonly achieved through natural regeneration. The postwar housing boom ended the Forest Service's role as a custodial manager and saw the onset of a period of active management with ever-increasing harvest levels. NFS harvests during the early postwar period grew dramatically.

By the early 1960s, many Americans became interested in environmentalism. At this time, increasing concerns were being expressed about the "other" forest outputs. While the NFS had been in its custodial phase, these forests had provided recreation, wildlife habitat, grazing, and other uses almost as an aside. In 1960, these "other" uses were codified in the Multiple-Use Sustained Yield (MUSY) Act, which explicitly recognized a number of additional multiple uses, such as recreation and wildlife, and instructed the Forest Service to manage for all of these uses.

When concerns over the "mix of outputs" continued to result in substantial conflicts and litigation in the 1970s, new legislation was proposed. The Resources Planning Act (RPA) in 1973 and the National Forest Management Act (NFMA) in 1976 directed the Forest Service to manage for multiple uses through a forest planning process that would demonstrate that all the outputs had been given a fair hearing. Furthermore, the Forest Service was instructed to develop a process to include public participation. It was believed that by involving all the interests in the process through planning and public participation, the differences could be resolved. Congress also indicated that it would sweeten the pot by appropriating additional funds so as to provide a bit more for all the forest interests (at taxpayer expense, of course). At the time, Senator Hubert Humphrey suggested that a forest plan should not cover more than 20 pages.

Forest Planning Problems

Forest planning, as practiced by the Forest Service, has largely been a failure. Planning and public participation have not reduced the conflict, and the plans have not provided serious guidance for implementation. Instead, the planning process has been much longer, more expensive, and more contentious than anyone foresaw. Rather than the 20-page plans envisaged by Senator Humphrey, forest plans today cover hundreds of pages. Furthermore, the legislation has opened the way to endless appeals and litigation.

The experience of Forest Service planning is similar to that of large-scale long-term planning elsewhere in the world. Detailed planning is costly, highly inflexible, and usually outdated before it is completed. In practice, the plan may be ignored, as is common in the Forest Service, or it may even cause harm by preventing the timely adjustments and adaptations that are always necessary in a dynamic context. In an ever-changing world, a forest plan needs to be flexible and quickly adaptable, rather than massive, overly comprehensive, and rigid.

> *The planning process has been much longer, more expensive, and more contentious than anyone foresaw. Rather than the 20-page plans envisaged by Senator Humphrey, forest plans today cover hundreds of pages.*

Over the years, Forest Service legislation has gradually provided for the production of other uses. However, it does not provide for definitive guidance as to the mix of outputs. This has set the stage for continuing battles, traditionally between timber production and environmental outputs, but more recently involving recreation, grazing, and other uses. Legal codification provided the base that subsequently gave rise to a morass of litigation, which continues to this day as the various interests compete for a larger share of the pie. The multiple-use concept had been for a shared management of the forest resource by all the users. But the tendency has been for each of the various interests to try to capture the whole. Despite planning, the question still is, who should dictate what mix of outputs these forests should produce? That question remains unanswered by the current planning system.

When completed, forest plans regularly are contested, first through an appeals process, which fortunately has been modestly revised, and then often through litigation. Even when plans are finalized and avoid or overcome appeals and litigation, they often are not implemented because of the lack of a budget or because of administrative decisions related to changing circumstances or objectives. In fact, the Forest Service budget largely is prepared independently of the forest plans, so the budget rarely comports well with the plans, and funding for their full implementation is rarely available. Also, administrative action, such as the recent roadless area initiative, supersedes all forest plans, thereby making many of them moot.

Other legislation, such as critical habitat set-aside associated with the Endangered Species Act, often negates plans as well. For example, timber harvests in the Southwest in recent years generally have failed to approach their targeted levels, but they almost surely will decline further as a result of the creation of large areas of critical habitat for the Mexican spotted owl. Not that this is necessarily a bad thing. But it demonstrates that the forest plans for this region, meticulously developed at large costs, will never be implemented and, in fact, have contributed little to the long-run management of these forests. Situations such as this one are common throughout the NFS. Other events continually render forest plans obsolete, often before they are fully completed. Thus despite the large amount of both public and private resources involved in their preparation, the forest

> *Events continually render forest plans obsolete, often before they are fully completed.*

plans have neither provided a basis for actual implementation nor reduced interest group contentiousness.

Recommendations

The new streamlined planning process that I recommend would articulate overall objectives and use a zoning approach to determine management activities appropriate to an area. But instead of a costly and onerous Environmental Impact Statement (EIS) or Environmental Impact Assessment (EIA), the process would use an Environmental Management Systems (EMS) approach. Such an approach could be consistent with the National Environment Protection Act (NEPA) by the addition, within the NEPA regulations, of a simple acceptance of the validity of the EMS approach. The EMS approach, similar to that of the International Standards Organization (ISO) commonly used by the private sector, certifies that operating management systems have the capacity to adequately monitor, update, and verify what is happening on the ground. The planning process would continue to use public participation, one of the strengths of the existing process.

> *The Environmental Management Systems approach certifies that operating management systems have the capacity to adequately monitor, update, and verify what is happening on the ground.*

This revised process would have the virtue of saving very substantial amounts of Forest Service resources, perhaps 40 percent of the resources now going to planning, which could be put to better use in managing and protecting the forests. In addition, private resources could be saved by reducing the onerous detail that public participants must master and over which they may become contentious. Furthermore, this more flexible approach would help the Forest Service regain credibility as an agency that is able to implement plans and promises in a manner consistent with changes that are occurring in the world.

> *Perhaps 40 percent of the resources now going to planning could be put to better use in managing and protecting the forests.*

I do not claim that using a radically simplified and less detailed planning system will eliminate the existing contentions associated with forest management. However, the planning costs can be dramatically reduced, by up to 40 percent, by providing a short, overall, general plan with minimal detail and maximum flexibility. Also, by relying on an EMS approach to ensure the monitoring of performance, rather than the torturous EIS or EIA, excessive amounts of data collection and litigation costs can be avoided. Such a process can provide for public participation without addressing a host of hypothetical management decisions that often never transpire because of the changing situation.

Conclusions

The extensive planning process of the Forest Service is a failure. It has become extremely costly, consuming hundreds of millions of dollars of appropriations, but it provides little in return. Despite the huge expenditure of resources, the planning process has not achieved the purposes for which it was created. Intended to provide brief general overall guidelines to the future management of the forests, it has burgeoned into a massive, unmanageable process. The contentiousness associated with Forest Service management, which was intended to be resolved through planning, has not abated, and litigation and appeals continue to flourish. Even when a forest plan successfully navigates the bureaucratic and legal systems, it is rarely implemented; a lack of funds, changing circumstances in the forest, or new overriding policies and directives bring it to a halt. The resources directed toward planning could better be saved by the taxpayer or redirected to more pressing needs.

Many of the conflicts between large-scale timber production and environmental goals, which provided the initial rationale for forest planning, have disappeared as Forest Service timber harvests have declined substantially. Thus some of the issues that the current planning system was designed to address are no longer relevant. While the Forest Service still faces substantial issues related to its management of the NFS, the complex and costly planning system currently in use has contributed little to resolving these issues.

The planning process should be dramatically revised and simplified in the spirit with which the legislation was created. Planning can be made far less costly by streamlining the process, eliminating the great amounts of detail and long-term projections, and substituting an EMS for the burdensome EIS and EIA approaches.

R.A.S.

Smarter Budgeting for Space Missions

by Molly K. Macauley

Mr. President, I strongly recommend that your administration reform the method by which the National Aeronautics and Space Administration (NASA) budget is allocated and administered. Doing so will avoid the massive cost overruns that plague the space station, space shuttle, Hubble telescope, and virtually every other ongoing program. Unless corrected, this problem will undermine all future NASA projects as well. The new approach that I propose—known as "incentive-compatible" institutional design—will accomplish two challenging objectives: it will minimize wasteful spending and, at the same time, enable NASA to better manage the highly uncertain costs naturally associated with space missions that push the frontier of technological know-how.

Background

Cost overruns have caused the space station and shuttle programs virtually to shut down—an embarrassingly detrimental outcome for the space program and the United States. The space shuttle program faced a $1 billion shortfall even before the Columbia shuttle accident in 2003, and the international space station is $4.8 billion, or 40 percent, over budget. NASA has had to cut the space station budget, cancel a habitation module and crew-escape vehicle, and reduce the station's crew from seven to three.

One of the rationales that NASA often offers for its budgetary problems is that developing new space technology is inherently uncertain. As a result, the agency justifies perennial cost overruns. Reforming the NASA budget will both provide incentives to minimize cost overruns and help ensure that the agency carries out any space vision that you may have with financial and technical success.

At present, a built-in incentive exists for wasteful spending because of the way the NASA budget is distributed within the agency. Contrary to public impression, the bulk of the nation's space activities—and the bulk of our $15 billion public expenditure on space each year—takes place not at NASA headquarters in Washington, but at NASA centers and facilities located across the country (see Figure 19-1).

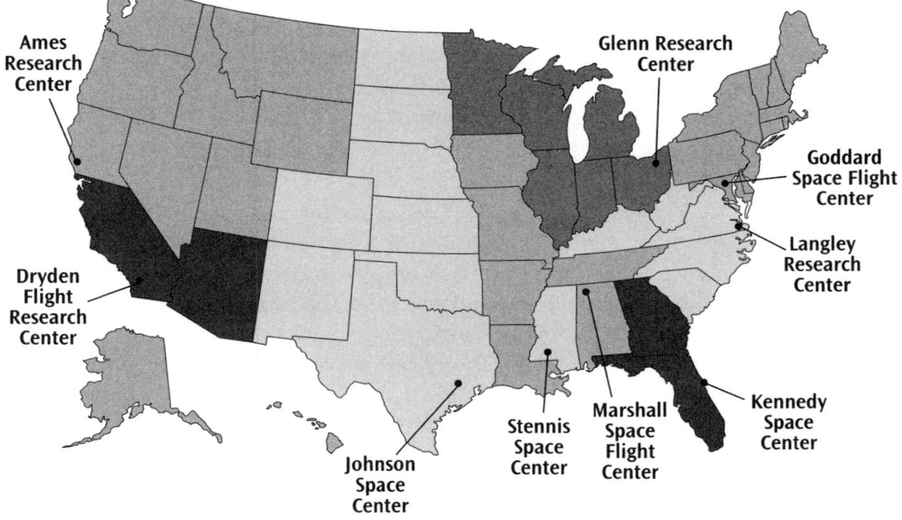

Figure 19-1

NASA facilities and the regions they serve.

Ames Research Center

Glenn Research Center

Goddard Space Flight Center

Langley Research Center

Dryden Flight Research Center

Kennedy Space Center

Johnson Space Center

Stennis Space Center

Marshall Space Flight Center

Some observers have suggested that it is time for an independent commission to review these facilities, much as the Department of Defense did in the process of military base closures and realignment that began in the late 1980s and continues today. The 2004 Presidential Commission on Moon, Mars, and Beyond recommends that NASA facilities be reorganized as federally funded research and development centers (FFRDCs). A review of NASA's facilities could lead to a variety of outcomes, including closing some facilities outright, combining the functions of different NASA centers, establishing the centers as FFRDCs, redirecting a specific mission or function of a center, or even forming public-private partnerships to lease out underused facilities.

In short, there is wide leeway for a creative, fresh look. But even if some of these steps were taken, and even if some centers were closed, a key problem will still loom large at NASA facilities: how best to manage the highly uncertain costs associated with some missions that push the frontier of technological know-how—the kind of research and development that is unlikely to be carried out profitably by the private sector. Development and testing of one-of-a-kind instruments, spacecraft, and propulsion and launch systems can be expensive, be technologically risky, and result in unmarketable products. But this fact should not result in your giving a blank check to NASA centers that conduct critical R&D.

How are you, as well as the NASA administrator, the Office of Management and Budget, Congress, and the public, to know what a new technology, fraught with unknowns, should cost? And why are there almost always cost overruns? Certainly underruns are possible, but the present system discourages them, because it gives the NASA centers no incentive to save. As with most government agencies, the budget process takes away savings rather than rewards them.

A key problem will remain: how best to manage the highly uncertain costs associated with missions that push the frontier of technological know-how.

Recommendations

The best solution is one studied by space policy analysts, an approach specifically tailored to the problem NASA faces. The innovation here is an entirely new relationship between NASA headquarters and its centers in managing the cost of space projects. At present, the centers carry out missions subject to costs that are capped at a budget negotiated early in mission design and planning. These caps pose severe problems. Projects can encounter unexpected design, fabrication, testing, and integration troubles. And the cap may not give a manager the leeway needed to innovate in order to solve the problem. In addition, managers admit that they avoid taking the risks inherent in innovation, preferring to stick too closely with reliable, proven technology. These safe bets may help mission launch and deployment, but sticking with old technology may well reduce the returns to science—the new information, data, and insights that could be gained from a more daring approach that truly pushes the space frontier.

> *Managers admit that they avoid taking the risks inherent in innovation, preferring to stick too closely with reliable, proven technology.*

The caps work the other way, too. They provide little incentive to save money if any appropriate cost reductions occur to engineers while carrying out the mission. Many managers report that they hurry to spend their budgets as the end of the fiscal year approaches, lest they "lose" the money.

These problems could be avoided, or at least reduced, if the centers had a pool of funds they could draw from for overruns or add to if managers and engineers find ways to save costs; and in this case, cost savings would not be taken away. Centers would have ownership and discretion for use of the funds, provided they maintain solvency and are subject to some general rules governing oversight. The rules could permit centers to use some of the funds for certain types of other investments, too, such as infrastructure or staff training.

To illustrate how this "bank account" approach might work, suppose that a NASA center is to design, build, test, and then launch a new rover for Mars within the next two years. The engineers face numerous engineering trade-offs: how much automation to build onboard the spacecraft versus relying on commands from controllers on the ground; how much onboard electrical power to provide, which ultimately determines how long the spacecraft can operate; and how much new technology to try out versus relying on old, off-the-shelf technology. Given the fast pace of technological innovations in lightweight high-strength materials, new battery designs, and new robotics techniques, engineers may find new ways of accomplishing the project during its two years of construction, and some of them are likely to save money. The incentive to take advantage of such savings is absent under the present system, but if the center can keep any savings, reallocating them to the next project or to other activities such as upgrading the center's infrastructure or offering employee rewards, this would give engineers a much stronger motivation to make trade-offs that make sense. The account approach also is likely to balance the tendency for managers to understate their initial estimates

> *A "bank account" approach would allow the center to keep any savings, reallocating them to the next project or to other activities such as upgrading infrastructure or offering employee rewards.*

of project costs, a practice that often causes the White House to be grossly misinformed about what a project is likely to cost.

Would center "bank accounts" work? Researchers at Caltech, Purdue, and other institutions have tested the idea in hypothetical mission design projects involving real NASA managers. Compared with business as usual, the bank account approach worked well. Bank accounts led to less cost escalation during the life of the project, reduced mission costs overall, resulted in fewer delayed missions, brought equal if not greater scientific return, and encouraged more innovation. The approach gets financial incentives right, much in the same way that the U.S. Federal Communications Commission has successfully auctioned licenses to use some regions of the electromagnetic spectrum. Those auctions brought new opportunities for innovative telecommunications products and low prices for consumers. The approach also is similar to the now routine buying and selling of pollution permits by electric utilities under the oversight of the Environmental Protection Agency.

Conclusions

A significant step separates experimental settings from the real-world operation of an agency. But countless examples in the public and private sectors show that financial incentives do work. They can and should play a much greater role at NASA. Because the approach deals with the budget, it may well require congressional approval. And like any major reform, it will take time and leadership to achieve acceptance and implement a new way of doing business. But alternatives such as center closing and realignment also would be radical reform and likely unpopular. Please consider banking on the centers to do their job.

M.K.M.

CHAPTER
20

Getting Serious About Antibiotic Resistance

by Ramanan Laxminarayan

Every year in America, some 40,000 people die from infections caused by antibiotic resistant bacteria, only slightly less than the number of those killed on our roads. This may actually be an underestimate, because many other deaths, particularly those of elderly patients suffering from a myriad of problems, may in reality be caused by these so-called superbugs. Mr. President, in order to address this serious threat to the nation's public health, I recommend that you introduce legislation to discourage the use of antibiotics where they bring little benefit to patients. These measures would include increasing cost-sharing for antibiotics under Medicare and Medicaid, while expanding coverage and subsidies for cough and cold medications that patients may use instead of antibiotics. In addition, a large proportion of antibiotic use in the United States is by the animal industry to help farm animals gain weight faster. Studies indicate that the benefit of using antibiotics for this purpose is greatly outweighed by the cost imposed on the rest of society in terms of reduced antibiotic effectiveness. In light of this, I also propose a ban on the use in growth promotion of all antibiotics that are currently being used to treat human infections.

The Problem

Modern medicine rests on the bedrock of the availability of affordable and effective antibiotics. In recent years, bacteria have been increasingly resistant to antibiotics, threatening our ability to treat previously treatable illnesses. Reports of methicillin-resistant *Staphylococcus aureus* and penicillin-resistant *Streptococcus pneumoniae* are increasingly common, to name just two dangerous pathogens. In fact, the prevalence of high-level penicillin resistance in *S. pneumoniae* in the United States grew 800-fold between 1987 and 1999, from 0.02 to 16.5 percent. According to the U.S. Food and Drug Administration (FDA), "Unless antibiotic resistance problems are detected as they emerge, and actions are taken to contain them, the world could be faced with previously treatable diseases that have again become untreatable, as in the days before antibiotics were developed." While the nation's focus has remained steadily on bioterrorist threats, a potentially greater threat already is lurking in our hospitals and communities and requires immediate attention.

Antibiotics have been in use in medicine since the 1940s and are now widely used for a variety of reasons, ranging from ear infections in children to keeping patients undergoing transplants free of infection. Although resistance has been increasing for a number of reasons, the underlying driving force is largely that doctors and patients who overuse, and sometimes misuse, antibiotics have no incentive to take into consideration the impact of their actions on the rest of society. This results in the use of antibiotics to treat viral and other conditions where they have no effect (antibiotics cure bacterial infections only), inadequate compliance by patients with treatment regimens, poor dosing by doctors, and lack of infection control, all of which have led to decreasing drug effectiveness.

And antibiotics are not only used in treating humans, but they also are used in livestock to help them gain weight faster (subtherapeutic use) and to avoid and treat disease (prophylactic and therapeutic use). It is estimated that more than half the antibiotics produced in the United States are used in the animal health industry, and the bulk of this is for growth promotion. Because these uses promote the development of drug-resistant bacteria in animals, and routes exist for the movement of these resistant bacteria to humans, drug resistance in bacteria associated with food animals can influence the level of resistance in bacteria that cause human diseases. Several compelling studies have documented the impact of subtherapeutic use of antibiotics on resistance in humans, and the evidence is mounting. Population biology predicts that the strong selection pressure imposed by the use of antibiotics in animal feed will lead to the evolution of resistant microorganisms. Although more studies will help improve our understanding of the links between antibiotic use in animals and resistant infections in humans, our wait for even more conclusive evidence will come at the cost of losing valuable drugs that will be very expensive to replace.

According to the U.S. Food and Drug Administration, "Unless antibiotic resistance problems are detected as they emerge, and actions are taken to contain them, the world could be faced with previously treatable diseases that have again become untreatable, as in the days before antibiotics were developed."

Recommendations

I recommend that you consider making two policy changes. The first is to discourage inappropriate antibiotic use with a combination of price and nonprice measures. Price measures would include subsidies or expanding Medicaid and Medicare coverage for over-the-counter substitutes for antibiotics, such as cough medicines and pain relievers. They also would increase copayments for antibiotics prescribed for conditions for which their use would bring little benefit to patients but would harm society by increasing drug selection pressure. Such price measures need to be combined with nonprice measures that include patient and physician education, better resistance surveillance data, increasing antibiotic heterogeneity, and providing warning labels on antibiotics. These nonprice measures alone are likely to be ineffective without a compelling economic incentive for patients and physicians. They must shoulder the cost that they impose on the rest of society in the form of resistance when they overuse or misuse antibiotics.

Second, I propose that you ban the use for animal growth promotion of antibiotics that are used in treating humans, unless the pharmaceutical companies can demonstrate that their use in growth promotion has no demonstrable impact on the evolution of resistant pathogens.

Raising the Costs of Antibiotics

The most reliable axiom in economics is that as the price of any commodity goes up, the quantity of that commodity that people will consume declines, all else being equal. Therefore, the most reliable way of reducing the use of antibiotics without second-guessing physicians' decisionmaking is by raising their costs to patients.

One solution might be to impose a tax on antibiotics, but this may be undesirable for two reasons. First, a tax may not discourage antibiotic use if insurance coverage shields many patients from drug costs and physicians are relatively insensitive to drug costs. Second, the burden of a tax may be borne disproportionately by poorer patients, who are less likely to have health insurance to cover the cost of antibiotic prescriptions.

A logical alternative would be to mandate an increase in the extent of cost-sharing for antibiotics. This could be accomplished by increasing copayments for antibiotic prescriptions for certain conditions for which a regulatory or scientific body believes that antibiotics are overprescribed, such as for the treatment of ear infections. Such a measure would not hurt the majority of economically disadvantaged patients who lack prescription drug coverage, but it still would effectively tax antibiotic use.

To be sure, a price-based policy intervention is a blunt instrument, and it may, in some instances, discourage the use of antibiotics even when their use is justified. However, targeted cost-sharing efforts aimed at certain diagnoses may be preferable to an across-the-board increase in mandatory cost-sharing for all antibiotics. Increased cost-sharing or other methods of raising the costs of antibiotics to patients may not be popular. But short of direct case-by-case oversight of prescriptions, few other alternative strategies could effectively lower antibiotic use.

Banning Use in Growth Promotion

Under a measure banning the use of antibiotics for growth promotion in animals, their use to treat sick animals would still be permitted. Although this policy change may raise the cost of meat production in the short term, and perhaps even increase the volume of antibiotics used to treat sick animals in the long term, withdrawing antibiotics as an input in the animal industry can bring great benefits to public health and safety.

Some countries in Europe already have adopted such a ban, and the lessons from their experiences can help guide your administration's decisionmaking. In Denmark, a ban on antibiotic use in swine production significantly lowered the use of antibiotics in growth promotion and raised the cost of swine production by less than 1 percent. Although it appears that the ban may have resulted in a higher incidence of disease among swine and a greater use of antibiotics used to treat sick animals, this may be just a short-term effect. The Danish farms that had the greatest difficulty coping with the withdrawal of antibiotics were those that were older and had poorer hygiene and less sanitary conditions. The use of antibiotics should not be a substitute for more modern methods of hygiene. A similar increase was noted in Sweden, where antibiotic use in animals was banned in 1986, but there this was a temporary problem. Over the longer term, livestock producers were able to move to an effective production system with lower antibiotic use.

The use of antibiotics should not be a substitute for more modern methods of hygiene in the food animal industry.

International consensus is growing on the need to safeguard our arsenal of antibiotics. The World Health Organization recommends that antimicrobials normally prescribed for humans should no longer be used to promote growth in animals. The European Union plans to phase out such use before 2006. The FDA already has recommended a withdrawal of the use of one class of antibiotics, fluoroquinolones, for use as growth promoters. These powerful antibiotics, which include the drug Ciprofloxacin, are valuable in treating people.

Conclusions

The problem of antibiotic resistance is of great concern to many federal agencies, although no single agency has a mandate to safeguard the effectiveness of these drugs. An interagency task force currently is looking into educational measures targeted at physicians and patients to reduce antibiotic prescribing, among other alternatives, to address the problem of drug resistance. However, without specific measures to create incentives for physicians, patients, and managed-care organizations to change their practices, these educational measures will not go far enough in curbing inappropriate antibiotic use.

For all of these reasons, I recommend that you support the bill introduced by Senators Edward M. Kennedy and Jack Reed, Preservation of Antibiotics for Human Treatment Act of 2002. Additionally, I propose that you expand this bill to include not just the use of antibiotics in animals, but also measures that address the use of antibiotics in humans. Finally, I urge you to act expediently. Our history of using antibiotics is only 60 years old, and we already are in danger of relying on using our most powerful drugs to treat humans. Increasing resistance may be irreversible, but managing these drugs as valuable societal resources may extend their usefulness to future generations.

One last consideration is that antibiotics are among our most valuable assets in defending against bioterrorism. Stockpiling antibiotics in case of an emergency does no good if these antibiotics are rendered ineffective by improper use. Without a farsighted, comprehensive agenda for dealing with drug resistance and coming up with imaginative solutions to a problem that is growing steadily worse, our ability to control infectious diseases is in danger.

Antibiotics are among our most valuable assets in defending against bioterrorism.

R.L.

Zoning the Oceans

Changing the Focus of
U.S. Fisheries Management

by James N. Sanchirico

O ur nation has one of the most productive and diverse marine environments in the world, but many of these resources are threatened. Mr. President, this does not have to be the case. Two commissions recently released reports detailing the causes, tools, and organizational restructuring needed to reverse this trend, both of which discuss the potential role for zoning the oceans. Sir, I recommend that you issue an executive order directing the secretaries of commerce and interior to launch a major new cross-agency initiative to zone our oceans by 2010, along with providing the necessary funding. Zoning the oceans would delineate portions of our marine environment for particular uses, just as on land. This is a bold idea that—unlike current approaches to management—can ensure economic growth and biological sustainability.

Background

The United States has about 95,000 miles of coastline and more than 3.4 million square miles of ocean within our exclusive economic zone. Within these waters exists a marine biodiversity rivaling that of some of the most treasured land-based systems. With habitats ranging from coral reefs and sea grass beds to salt marshes and mangrove forests, the ocean and coastal areas surrounding the United States and its territories are teeming with marine life.

These habitats and the resources they support provide the public with a valuable and diverse set of goods and services, including seafood, recreational enjoyment, carbon sequestration, storm protection, and opportunities for pharmaceutical discoveries. The National Marine Fisheries Service (NMFS) estimates that commercial fisheries alone add approximately $27 billion per year to U.S. gross domestic product. In addition, about 3.5 million acres of coastal wetlands provide many vital services and resources, including nursery, feeding, breeding, and resting areas for fish, shrimp, crabs, mollusks, and birds. The coastal environment also supports many tourism and recreation activities.

The ocean is critical to society's economic and social well-being, but scientific studies confirm that many of our marine resources are overexploited and face external environmental threats. NMFS reports that 76 of the 894 federally managed fish stocks are overfished and another 60 are subject to overfishing. Too many boats chasing too few fish bring our ocean resources to the brink of collapse and sometimes beyond, as happened in the New England cod fishery in the early 1990s and more recently in the West Coast rockfish fishery.

> *The ocean is critical to society's economic and social well-being, but scientific studies confirm that many of our marine resources are overexploited and face external environmental threats.*

Managing Living Marine Resources

Four concurrent paradigmatic shifts are under way in how societies approach managing marine resources. The first two, marine biodiversity conservation and the ecosystem approach, address the goals of management; the others, marine reserves and rights-based fishing, are policies that can reverse current trends and also address the expanded sets of goals. The underpinnings of these changes are discussed in the June 2003 Pew Oceans Commission report and the April 2004 U.S. Commission on Ocean Policy draft report.

Marine Biodiversity Conservation

A marine biodiversity conservation ethic is blossoming that will challenge the current management framework, which was designed for maximizing the returns from extractive uses. The constituent base for marine resources must get larger and more knowledgeable and vocal, however, before biodiversity conservation will be considered a legitimate goal in the political economy of marine policy. As it stands, according to a survey by the Ocean Project, the American Public possesses "only superficial knowledge of the oceans, their functions, and their connection to human well-being."

That the public is not fully aware of the oceans' importance is not surprising, as much remains that marine scientists do not yet understand. For example, the U.S. National Oceanic and Atmospheric Administration estimates that more than 95 percent of the world's oceans remains unexplored, and some estimate that about 5,000 marine fish species have yet to be discovered. Partially to address our knowledge gaps, the Census of Marine Life (CoML) project was launched in 2000. The project, funded by both public and private sources, includes some 300 scientists from 53 countries. Its goals are to describe marine biodiversity, to understand its role in the functioning of ocean ecosystems, and to standardize scientific sampling and data management protocols by 2010.

Although the American public does not currently list marine biodiversity as an issue of immediate concern, putting areas aside to conserve it would have real benefits. Many would argue that during the first years of the twentieth century Americans were not fully aware of the values and importance of terrestrial biodiversity for ecosystem functions. Yet President Theodore Roosevelt set aside places for perpetuity that today have become symbols of our nation.

Could coral reefs and seamounts become symbols of our national natural treasures over this coming century? Yes. It is also my belief that the world will witness in the coming years a movement to set aside areas of the ocean that will rival the U.S. land conservation movement of the early twentieth century.

Ecosystem Management

Fishery management has traditionally been designed one species at a time. Many implicate the single-species approach, scientific uncertainty, and tendencies to favor higher catch totals for sociopolitical purposes in our current crisis.

> *It is my belief that in the coming years a movement to set aside areas of the ocean will rival the U.S. land conservation movement of the early twentieth century.*

While many definitions exist on what constitutes ecosystem management, almost all are based on the idea that fishery management decisions should not adversely affect ecosystem function and productivity. This would entail setting policies to reduce habitat damage by mobile bottom gear, such as bottom trawling. It also would take into account the incidental catches of marine mammals by swordfish long-liners in the Pacific or sea turtles by shrimp fishers in the Gulf of Mexico. Additionally, fishery managers would need to take into account predator–prey relationships when setting catch levels to ensure that catches of one species do not affect the sustainability of the other species.

Recent proposed legislation in the 107th and 108th Congress emphasizes the need to design ecosystem-based management plans. But from an operational standpoint, many of the hard questions remain, such as what a "true" ecosystem management plan entails. In addition, it is not clear what ecological and economic trade-offs are inherent in an ecosystem plan. What is clear, however, is that an ecosystem approach is compatible with an integrated and well-planned zonal approach, as the instruments and uses in zones can incorporate their effects into ecosystem function.

Marine Reserves

Marine scientists are moving away from the assumptions that marine populations are evenly distributed toward notions of patchy habitats with population abundances varying across space. Because management historically has been characterized by systems of relative uniformity of regulatory actions over space, this shift enables fine-tuning that will lead to more biological and socioeconomic sustainable management.

Partly as a result of this new "patchy paradigm," in 2001 more than 150 marine scientists signed a scientific consensus statement outlining the current need for and benefits from the creation of marine reserves—areas of the ocean set aside from all extractive uses. The list of potential benefits includes conservation of biodiversity, sources of larvae, greater levels of biomass, increased catches, and a hedge against management failures.

Historically, very little of U.S. waters have been protected. However, largely at your discretion, Mr. President, that could change, as the Clinton administration issued two executive orders that put support behind the need for and use of marine protected areas, especially marine reserves. The first order created the Coral Reef Task Force, which recommends designating special areas, including "no-take zones, to protect and replenish coral reef ecosystems and prevent future harmful impacts." The second, and arguably more far-reaching, effort came in 2000, when federal agencies were instructed to develop a national system of marine protected areas (MPAs). The Marine Protected Areas Federal Advisory Committee, a multi-stakeholder group, was formed to advise the secretaries of commerce and interior on the implementation of a national system of MPAs.

Rights-Based Approaches

In order to address the problem of "too many boats chasing too few fish," and the economic and ecological waste of this phenomenon, managers need to implement rights-based approaches, such as individual fishing quotas (IFQs). Such quotas currently are used in many different fisheries and countries, including the United States, to regulate fishing efforts to the benefit of both fishermen and the environment. Another rights-based approach is a fishing cooperative in which fishermen are granted legal authority to collude to determine who will fish and for how much. Such a system was set up in the Alaskan pollock fishery in 1998 and is generally viewed as a success.

IFQ programs are analogous to other cap-and-trade programs, such as the sulfur dioxide allowance-trading program. They limit fishing operations by setting a total allowable catch, which is then allocated among fishing participants, typically based on historical catch. When fishermen have access to a guaranteed share of the catch, they have an incentive to stop competing to catch as much as possible and start improving the quality of their catch. When shares are transferable, inefficient vessels find it more profitable to sell their quotas than to fish them. The result will be fewer and more efficient vessels.

> *When fishermen have access to a guaranteed share of the catch, they have an incentive to stop competing to catch as much as possible and start improving the quality of their catch.*

Rights-based approaches are critical instruments that need to be included within a zonal management system for several reasons. First, in existing programs, governments charge the quota owners fees that recover the costs of management, such as data collection, scientific research, onboard observer programs, and other enforcement programs; this is very appealing, especially in this time of budget deficits. Second, IFQ programs offer a market-based solution to the overcapacity problem (addressing "too many boats") that does not rely on direct payments from the government in the form of vessel buyback programs. Finally, and most important, research at Resources for the Future has shown that the total value of New Zealand's IFQ fisheries, which account for more than 85 percent of the commercial catch taken in its waters, has more than doubled in real terms from 1990 to 2000, even as fish stocks are improving.

The Zonal Approach

Employing a zonal approach is beginning to be discussed in national and regional meetings on how to improve U.S. ocean governance. In general, the approach requires looking at the system not as one biological, legal, and economic homogenous unit, but as an interconnected system of heterogeneous spatial units—much as we do with management of terrestrial natural resources.

All of these fundamental paradigm shifts can be incorporated within a zonal approach, but this cannot be done by tweaking current approaches. With a zonal system in place, a no-take zone designed, for example, to protect a unique habitat or rare and special ecosystem assemblage could be abutted against a commercial fishery harvest zone employing individual fishing quotas.

If marine reserves are not coupled with approaches that address incentives, then any economic gains and most likely the biological gains will be dissipated as fishermen continue to race for the fish— the very circumstances that created momentum for increasing the scale and scope of marine reserves.

If marine reserves are not coupled with other approaches that address the incentives fishers face, then any economic gains and most likely the biological gains from a reserve will be dissipated as fishermen continue to race for the fish—the very circumstances that have created momentum for increasing the scale and scope of marine reserves. On the other hand, a zonal approach can ensure both the efficient utilization and sustainability of the ecosystem beyond the boundaries of the current proposed no-take zones, thereby improving the long-run health of the entire marine ecosystem.

In addition to accommodating marine reserves and IFQs, zones could be set up for recreational fishing, with some of these allowing only certain types of gear, similar to fly-fishing-only sections on trout streams, or for charter-fishing operations for certain species at certain times of the year. Near-shore areas could remain public-access zones to allow surf fishing, beach recreation, and other uses. Certain areas offshore also could remain open access, to address concerns about "fencing off the last frontier." Zones could be created for aquaculture, natural gas and oil operations, seabed mining, and offshore wind farms.

Rights to particular zones could be leased for certain uses for part or all of the year. Leases could be auctioned off to participants or grandfathered. Such auctions could generate an additional source of government revenue, just as the spectrum auctions have done. Fishing clubs or charter-fishing operations could purchase long-term leases that would guarantee their members or customers exclusive rights to certain areas. Offshore wind farms could join forces with offshore aquaculture operations to purchase the rights to certain areas. Environmental organizations could purchase leases to create private marine reserves that could complement government conservation efforts. And creating quasirights to areas could lead to the development of secondary lease markets. For example, an oil company might lease out the fishing rights in its zone, or a pharmaceutical company might purchase the rights for bioprospecting operations.

The number of uses of the marine environment is not going to decrease during your tenure, and as the number of stakeholder groups increase, so too will the need to address user conflicts. Zoning can address these conflicts in one of two ways:

either regulators could separate activities in a traditional command-and-control approach, or conflicts among the different stakeholder groups could be left to the lease markets.

Whatever the instruments used or types of uses allowed, a zoning system can be monitored and enforced with satellite tracking systems that either allow or lock out certain activities by various individuals at different times and places. Technology currently exists to "fence the ocean" electronically, if not literally.

A zonal approach can ensure both the efficient utilization and sustainability of the ecosystem.

Besides the many benefits in moving to a zonal approach, some transaction costs and uncertainties need to be considered. Changing the focus of current fishery management will require jurisdictional and institutional restructuring. Also, many marine species move great distances during their lives and will not stay within artificial lines "drawn" on the water. Flexibility in how zones are defined and managed, along with designing the system using the best available natural and social science information, is therefore critical for this—or for that matter, any—marine management framework to be successful.

Conclusions

In an important sense, we are on the cusp of a revolution in managing marine resources. Our marine resources do not need to be threatened. Fishing companies and the communities they support do not have to go the way of old-growth logging towns. Future generations can enjoy the bounty of the oceans much as we do today. Reversing the current trends, however, will not be easy—many will not get what they want—but zoning is likely to be the least costly and most ecologically sound approach.

J.N.S.

Part IV
Information Decision Frameworks

Combatting Ignorance About U.S. Water Quality

by James Boyd and Leonard Shabman

Gaps in basic information about water quality hobble attempts at all levels of government to improve the quality of our nation's water resources. Mr. President, a National Water Quality Monitoring Strategy, developed under your leadership and supported by changes to the Clean Water Act, can close these gaps and make possible more efficient and effective water quality management. Tracking national trends in water quality conditions requires a new multiagency effort. This effort must be authorized under the Clean Water Act and be appropriately funded. We recommend that, using amendments to the Clean Water Act, you seek authorization to increase appropriations by 10 percent per year over the next 10 years to support the states' responsibility for water quality assessment and management in their rivers, lakes, and estuaries. Additionally, we recommend that you initiate an interagency cooperative investment program for the research and development of new technologies to improve the quality and affordability of ambient water monitoring.

The Nation's Water Quality

Most Americans would be surprised to learn how little we know about the nation's water quality. For the nation as a whole, and in most areas, we do not know if water quality standards are being met or whether our water quality management programs are yielding improvements. When we find problems or improvements, we rarely have the data to explain the underlying causes. This knowledge gap is unacceptable.

The American public ranks water quality protection among its highest environmental priorities, and as a nation, we spend billions of dollars every year on the improvement of water conditions. Yet we are unable to reliably characterize the condition of many rivers, lakes, and estuaries. Numerous water protection programs have been put in place, but our major water quality statute is 32 years old.

Our ignorance comes at a cost. Serious undetected water quality problems may exist. Equally important, once problems are detected, limited information makes it impossible for localities, states, and federal authorities to identify the most cost-effective response. Then, when action is taken to address a water quality problem, we are unable to judge the policy or investment's performance. When it comes to the nation's waters, the highest unmet priorities are identification of the biggest problems and the ability to measure success and failure.

> *When it comes to the nation's waters, the highest unmet priorities are identification of the biggest problems and the ability to measure success and failure.*

The reason so little is known about water quality conditions is that we are embarking on a new era in water science and regulation. The national approach to water quality management in particular is undergoing a profound shift. In the first 32 years of the Clean Water Act (CWA), attention focused on limiting releases from industrial, commercial, and municipal point sources. Specifically, attention focused on writing, implementing, and enforcing a series of regulations requiring point sources to install mandated pollution control technologies. When the focus was on limiting and monitoring the releases from these point sources, little attention was given to the overall condition of lakes, rivers, and estuaries. Although significant reductions in many pollutants from point sources have been achieved, now the focus is shifting to the condition of waters themselves and the entire range of pollution sources, including agricultural and urban runoff. At the heart of this shift is the so-called Total Maximum Daily Load (TMDL) program. A long-neglected aspect of the CWA, TMDL provisions require states to identify waters that are not in compliance with water quality standards, establish priorities, and implement improvements—including improvements that rely on nonpoint-source reductions. This process places a premium on credible water quality information.

The TMDL process has taken center stage because of widespread suspicions that significant water quality problems remain. Where surveys are conducted, often with reliance on very limited data, states report that about one-third of stream miles and 40 percent of the area of lakes and estuaries may not be meeting water quality standards. Put differently, states report that 300,000 miles of river and 5 million acres of lakes are not clean enough to support swimming, fishing, or boating. The National Water-Quality Assessment (NAWQA) program of the U.S. Geological Survey (USGS), while not intended to be a comprehensive national monitoring network, reports that 20 percent of groundwater samples collected exceed drinking-water standards for nitrate concentration, 80 percent of streams sampled have concentrations of phosphorus greater than EPA goals for preventing nuisance plant growth, and pesticides are detectable in 95 percent of groundwater samples. Regulatory and investment needs for the TMDL program demand more intensive attention to localized areas where water quality problems are suspected. Currently, each state targets its monitoring programs differently in light of its own needs, and few can afford a reliable statewide sampling program. Accordingly, available monitoring data tend to not be representative of statewide conditions. When aggregated to the national level, current data are unlikely to be representative of nationwide conditions.

The limited quantity and reliability of water quality information has been recognized by, among others, the states, public utilities, industry, environmental groups, the General Accounting Office, and the National Research Council. According to the 2000 National Water Quality Inventory, two-thirds of the nation's water bodies are unassessed, meaning that sufficient data do not exist to evaluate water quality conditions in many areas. This year, the director of the Environmental Protection Agency's Water Office argued that "we are flying blind" when it comes to water quality data. Improving information about water quality conditions requires a monitoring program to collect enough reliable data, basic scientific understanding of how watersheds work, and the advancement of analytical capabilities necessary for interpretation of these data. Federal leadership to improve our national water quality information is both necessary and appropriate. The focused development of modern data collection methods and analytical tools for data analysis is an activity that serves all states and localities. Basic science research to understand how watersheds work will serve all states and localities. The collection and analysis of data on interstate rivers is a logical federal responsibility. A federal water quality information initiative will allow you to demonstrate your commitment to progress on an issue that plagues our environmental programs at all levels of government. It is a broad initiative that requires and is worthy of a president's attention.

Benefits and Challenges

Improved water quality information will prove valuable for several reasons. First, it will reduce the chance that significant problems are overlooked until we are surprised by human health or ecological injuries. An example of this is the so-called Gulf hypoxia problem. In the last decade, ocean scientists discovered and documented a vast expanse of water in the Gulf of Mexico with reduced oxygen levels, thought to be due in large part to agricultural runoff from the Mississippi, Missouri, and Ohio River basins. More comprehensive monitoring will allow such problems to be detected earlier. In turn, this will help minimize the ecological and economic impact of water quality degradation.

The economic value of better information should not be ignored. Water quality problems can affect large investments made by both the public and private sectors. The success of land development decisions, public infrastructure projects, and industrial production can depend on water quality conditions. Water quality problems can trigger legal issues that force limitations on investment, development, or production. In other cases, businesses rely directly on the availability of clean water. This is true of many industrial facilities, agricultural operations, and recreational service providers. When information on water quality is poor, investments are made on the basis of that poor information. As any investment analyst will tell you, bad information can lead to bad decisions and poor economic outcomes. Toward that end, a central goal of better information is the ability to manage for performance. Monitoring systems should be designed explicitly to deliver information that will allow localities, states, and the nation to experiment with—and judge the performance of—different approaches to water

> *The economic value of better information should not be ignored. Water quality problems can affect large investments made by both the public and private sectors.*

quality improvement. Public sector expenditures, such as agricultural conservation payments, and private sector costs, such as the costs of controlling pollution, can be lowered if we are better informed about what expenditures provide the biggest environmental benefit.

Second, improved information will improve the quality of regulation. U.S. water quality regulation already is moving toward watershed-based, ambient regulation. Improved information will vastly improve the quality of this regulation. The maxim that "what gets measured gets managed" should be amended to say that we "regulate even when we do not measure." Our current inability to measure the performance of private or public actions to improve water quality means that we may be wasting money on policies that have little benefit. Also, our inability to measure makes innovative approaches to regulation impractical. Group discharge permits and water quality trading, by allowing flexibility, could help reduce the costs of achieving water quality improvements. These policy innovations are not possible, however, until environmental performance is reliably measured. Similarly, better information will significantly increase the environmental improvements we get from farm program expenditures dedicated to conservation. When performance can be measured, performance can be demanded.

Third, better information lowers the cost of conflict. Waters are shared by a wide variety of public interests, including agriculture, commercial businesses, municipalities, and recreational enthusiasts. This means that water management decisions and policies often generate social conflict. And conflict is costly. A recent controversy involving the Missouri River is illustrative; here commercial interests seeking the ability to navigate large vessels up the river are pitted against outdoor groups and service industries that want enhanced habitat and water quality. Poor information exacerbates such conflicts. Better information will reduce uncertainty and enhance the credibility of analysis aimed at balancing competing social interests.

Poor information exacerbates conflicts. Better information will reduce uncertainty and enhance the credibility of analysis aimed at balancing competing social interests.

Responsibility for monitoring currently is shared across all levels of government. We believe that a federal leadership role in promoting a strategic approach to water quality information is needed, but this is hampered by law and budgetary circumstances. A challenge will be to settle on the monitoring and assessment responsibilities of each level of government. Stronger federal leadership in the provision of water quality information does not imply a change in responsibility for planning and executing water quality management programs. States and local governments will retain their core responsibilities in the regulation and management of water quality.

Finally, monitoring data—even if improved substantially—cannot describe the condition of any river, lake, or estuary with absolute certainty, especially in consideration of the reality of resource limitations that limit sampling scope and intensity. As a result, the design of a national monitoring strategy should accommodate and include communication strategies to report unavoidable margins of error. Ways to communicate margins of error to the public and to incorporate the reality of this error into decisionmaking must be part of the national water quality information system.

Recommendations

We propose that you appoint a multiagency intergovernmental task force on water quality monitoring. This immediate effort, on a six-month time line, should document the extent, scope, and consequences of current water quality information gaps and set priorities for the most important data needed to fill those gaps. This initial gap analysis is properly led by EPA but should emphasize cooperation and input from the full range of agencies and levels of government.

Funding for a national monitoring program must be steady and sustainable. Money is wasted when monitoring systems are put in place and then starved of future funding.

You should ask the task force to explicitly provide you with recommendations that address funding and the proper division of federal and state responsibilities. Funding for a national monitoring program must be steady and sustainable over the long term. Money is wasted when monitoring systems are put in place and then starved of future funding. Today's information gaps are due to limited funding levels and a lack of legislative clarity on the purposes to be served by monitoring programs.

We make several specific recommendations. First, you should ask the task force to propose amendments to the CWA to authorize a national program to assess water quality conditions and trends and track them consistently over time. The demand for national scale assessment information is principally a federal responsibility, although a national program cannot be developed or implemented without the support of the states. Appropriations for this program should be $30 million per year, increase over time, and be in addition to existing monitoring funds now available to the agencies. Leadership should be by EPA, but a strong partnership must be fostered with the monitoring programs of other federal agencies, in particular the USGS.

Second, you should ask the task force to propose amendments to the CWA to increase appropriations to support state responsibilities for water quality assessment and management in their rivers, lakes, and estuaries. Funds should increase federal support for state monitoring programs by 10 percent per year for 10 years. Appropriations should be made as grants to the states based on a formula considering population, stream miles and lake and estuarine acres, and state cost-share offers. The task force should explore ways in which the federal government could support state efforts to place fees on water discharge permit holders, dedicated to meeting the cost-sharing requirements of a new federal monitoring grants program. Existing funds also should be deployed in the most cost-effective ways. For example, opportunities to shift monitoring locations and protocols on regulated sources to support watershed scale assessment should be identified. This might be done as part of EPA's oversight of the National Pollutant Discharge Elimination System (NPDES) program.

Finally, we suggest that the task force give you recommendations for a new federal interagency cooperative R&D investment program—including the National Science Foundation (NSF), EPA, NASA, and USGS—for the development of new technologies to improve the quality and affordability of ambient water monitoring. Monitoring costs are due primarily to the costs of data collection and transmission.

Emerging technologies that feature rapid advances in remote sensing, continuous monitoring instruments, and real-time data transmission methods promise to significantly reduce these costs. Unmanned, continuous monitoring with automated data export is already a reality for some water quality measurements and can play a key role in a lower-cost, comprehensive monitoring system. But R&D in this area is not coordinated or funded in a strategic manner. A federal R&D strategy is justified by economies of scale in monitoring research and technology transfer. Also, EPA should be instructed to standardize the procedures by which monitoring technologies are certified. This will enhance the rewards to innovation and promote the timely introduction of superior technologies.

J.B.
L.S.

Create a Bureau of Environmental Statistics

by H. Spencer Banzhaf

The importance of collecting economic statistics to track the progress of economic growth and to guide fiscal and monetary policy has long been recognized. As environmental policy grows in importance, the need for a similar set of comprehensive, consistent statistics on the quality of the environment likewise is growing. Mr. President, to meet this need, I recommend that you call for the creation of a new statistical agency—a Bureau of Environmental Statistics (BES), modeled after success stories such as the Census Bureau, the Bureau of Labor Statistics, the Bureau of Economic Analysis, and the Energy Information Administration. A fully funded BES would be able to expand the set of data collection tasks to meet policy requirements, and to centralize and coordinate current activities.

Rationale for a BES

Why do we need another bureaucratic agency collecting statistics? The overarching reason is that we simply do not have an adequate understanding of the state of our environment. In many cases, the network of monitors measuring environmental quality is insufficient in geographic scope. Often our knowledge of national air quality is based on a few monitors per state; our knowledge of water quality is even weaker. The measures we do have typically focus on potential problem areas—a sensible approach from the standpoint of enforcement, but not for surveying the overall state of the environment. Accordingly, we must make inferences about overall quality from observations at these trouble spots. The consequence is a biased understanding of environmental quality. Of course, this easy answer begs the further question of why we need a better understanding of the state of our environment. There are several good reasons.

First, we have a natural desire to understand broad trends that affect our society and its welfare. Indeed, it is for this reason that we first began to collect many of our national economic statistics, including the familiar measures of gross domestic product (GDP) and inflation. Yet from the origins of GDP accounting, it was acknowledged that GDP is only a proxy and not a perfect measure of welfare, because it omits many important components that do not pass through markets. Even then, the environment was acknowledged to be one of the important omissions. Since that time, we have invested enormous resources in improving measures of the market components of national well-being, but we have not sufficiently broadened that effort to other components, such as the environment. It is time to do so.

Second, our ability to design effective policies to balance environmental quality with other objectives, or to attain environmental objectives in the most efficient and effective manner, is hampered by inadequate information. Looking in the rearview mirror, in many cases we do not know whether existing policies have been effective, making it difficult to assess what remains to be done. Looking forward, we often find that the playbook of strategies with which we might attack environmental problems is limited for lack of information. Sometimes the lack of information creates practical problems for implementing and enforcing a strategy. For example, it is difficult to imagine a serious effort to manage the total maximum daily load (TMDL) of pollutants into our nation's watersheds, as the U.S. Environmental Protection Agency (EPA) has proposed, without more complete data about pollution loadings and their sources. At other times, the lack of information makes it difficult to anticipate the effects of a policy, creating political uncertainties. For example, the cap-and-trade system, proven a highly cost-effective way to reduce air pollution nationally, may allow remaining pollution to concentrate in particular areas. Without a more thorough monitoring network, it is impossible to know whether these so-called hot spots are a serious problem. The consequence is hesitation in further use of this potentially effective policy instrument.

We currently spend some $150 billion annually on measures to comply with environmental regulations, without a commensurate effort to understand the return from these expenditures.

A third reason we should want better environmental statistics is that many expensive environmental regulations, with serious consequences for businesses and local economies, are triggered by incomplete information. We currently spend some $150 billion annually on measures to comply with environmental regulations, without a commensurate effort to understand the return from these expenditures. Moreover, the indirect effects of compliance can be severe in some regions. A prominent example is compliance with air quality standards: counties and regions that fail to meet these standards risk loss of federal highway dollars, bans on industrial expansion, and mandatory installation of expensive pollution-abatement equipment. Compliance is often based on readings from a small number of monitors. A fair question is whether some communities have been singled out while others have escaped detection. Moreover, although readings from only one monitor may push a portion of a county over a pollution threshold, reestablishing a clean slate once air quality has improved is much more difficult. The catch-22 is that a county must prove compliance throughout its jurisdiction, even if the monitoring network is inadequate to shed light on all areas.

Creating a BES also would facilitate "one-source shopping" for members of Congress, agency administrators, and the public, who currently must navigate a maze of agencies to construct a picture of the nation's environment. In addition, an independent BES might lend more credibility—a sense of objectivity—to our environmental statistics, giving the public square a commonly accepted set of facts from which to debate policy, much as the Bureau of Labor Statistics (BLS) and the Bureau of Economic Analysis (BEA) have done for economic statistics, the Bureau of Census (CEN) for demography, and the Energy Information Administration (EIA) for energy.

Applying Earlier Lessons

Our experience with economic statistics teaches us a number of lessons for a BES. First, statistics can be politically controversial. Although widely accepted now, some economic statistics were the focus of past controversy. During World War II, for example, industrial wages were linked to changes in the U.S. Consumer Price Index (CPI). At the same time, the CPI began to move out of sync with the popular perception of price changes, recording much lower inflation rates than people were experiencing in their everyday lives, largely because it missed quality deterioration in the goods selling at modestly increasing prices: eggs were smaller, housing rental payments no longer included maintenance, tires wore out sooner, and so forth. The result was political uproar, with protests on the home front from organized labor. In the end, a lengthy review process, with representatives from labor, industry, government, and academic economists, resolved the issue.

Although environmental statistics probably will never hit people's pocketbooks as directly as did the CPI, they can get caught in the crossfire between business and environmental groups. Building in a regular external review process would help keep the peace during such moments. Crises aside, external reviews would ensure that a BES is balanced and objective, in both fact and perception, and help improve its quality over time.

Crises aside, external reviews would ensure that a Bureau of Environmental Statistics is balanced and objective, in both fact and perception, and help improve its quality over time.

Indeed, the regular external reviews of the CPI have raised points that would be of value to a future BES. Some are academic questions about sampling and analyzing data and could be addressed within the agency. Others may require congressional action from the beginning, such as the need for data sharing. In our economic statistics, substantial overlap exists between information collected for the U.S. Census (housed within the Department of Commerce), the unemployment statistics and the CPI (collected by the BLS), and the GDP (collected by the BEA). To address this concern, Congress recently passed the Confidential Information Protection and Statistical Efficiency Act, which allows the three agencies to share data and even coordinate their data collection.

Similar data-sharing issues would arise for environmental statistics. Currently, environmental statistics are collected not only by EPA, but also by the Departments of Agriculture, Interior, Energy, and Defense. A major advantage of a BES would be the consolidation of many of these activities. Nevertheless, data-gathering activities across other agencies touching on the environment probably always would overlap to some degree, and so even would some economic data collection. Coordination across these agencies would be essential for creating the best product without duplication of effort.

An additional insight gained from looking back on our experience is that economic statistics now play a much larger role in our economy and in economic planning than originally envisioned. Most generally, they have been used as a scorecard for the nation's well-being, a basis on which leaders can set broad policy priorities (stop inflation, spur growth)—and a basis by which the public can assess its leaders. At a more detailed level, they now fit routinely into the Federal Reserve's fine-tuning of the economy. Finally, through indexing of wages and pensions, tax brackets, and so on, the CPI automatically adjusts many of the levers in the economic machine.

Environmental statistics eventually could play each of these roles. Despite their current weaknesses, they already help us keep score of our domestic welfare. And they increasingly could be used to adjust policies. Initially, they may serve as early warning signals for problems approaching on the horizon, or all-clear signals for problems overcome. Later, as the data develop and policies evolve to take advantage of them, they may even be used in fine-tuning: on theoretical drawing boards, economists already have designed mechanisms that, based on regularly collected data, would dynamically adjust caps for pollution levels or annual fish catches. The only thing missing is the data with which to make such mechanisms possible.

> *Initially, environmental statistics may serve as early warning signals for problems approaching on the horizon, or all-clear signals for problems overcome.*

A final lesson learned is that high-quality statistics cannot be collected on the cheap. We currently spend a combined $722 million annually on data collection for the U.S. Census (excluding special expenditures for the decennial census), the BLS, and the BEA, and more than $4 billion each year for statistical collection and analysis throughout the federal agencies. Over the past three years, these budgets have increased at annual rates of 6 to 10 percent. Nevertheless, these efforts are widely considered to be well worth the cost.

By comparison, the current budget of $168 million for environmental statistics seems small. Mr. President, again, consider that the annual private cost of pollution control is approximately $150 *billion*, and that government spends $500 million a year for environmental enforcement. With approximately 2 percent of our GDP at stake in these expenditures, and the welfare of many people, a top-notch set of environmental statistics seems long overdue.

Implementation Options

Unlike the birth of some of these earlier statistical agencies, which addressed new fields of study, the arrival of a new Bureau of Environmental Statistics would involve as much coordination and consolidation of existing activities as expansion. Consequently, it would without doubt stir up sibling rivalry among the existing agencies as they seek to defend their old privileges, or to be the heirs of the new ones. Accordingly, the bureaucratic location of the BES would be a real political issue. While the Environmental Protection Agency would make sense as the one agency exclusively devoted to protecting the environment, a case could be made for other arrangements as well. Ultimately, the most important thing will be the creation of the agency, not its political home.

Conclusions

With the greater scientific understanding today of ecosystems and their relation-ship to human welfare, and with the greater weight given to the environment in public policy, comes a greater need for environmental statistics to quantify those relationships and to guide policy. Likewise, as we become more conscious of the role of the environment as a factor in human welfare, we require more information about the state of the environment just to track how we are doing. Expanding the collection of environmental statistics and centralizing it under one roof—a well-funded Bureau of Environmental Statistics—would be the first step in this process.

H.S.B.

Treading Carefully With Environmental Information

by Thomas C. Beierle

Congress introduced information disclosure nearly two decades ago as a strategy for improving industrial environmental performance. Mr. President, I recommend that your administration reinvigorate environmental information disclosure as a policy tool. The effort could involve enhancing existing programs or creating an entirely new program. Your administration should evaluate whether a new disclosure program would help you achieve your top environmental goals. A new program might include, for example, a single point of entry for information on drinking-water contaminants or an integrated database on polluted coastal waters. The process by which a new or enhanced disclosure program is developed is as important as its substance. Policy development should include an open and transparent process for cataloging and rigorously evaluating the claims of proponents and critics of disclosure. Particularly important is that your administration seriously address the regulated communities' concerns about cost, privacy, and security through thoughtful program design.

Background

In 1986, Congress ushered in a new era of federal environmental management by introducing information disclosure as a strategy for improving industrial environmental performance. The Emergency Planning and Community Right-to-Know Act (EPCRA), passed that year, included a low-profile provision called the Toxics Release Inventory (TRI), which called on the U.S. Environmental Protection Agency (EPA) to make data on releases and transfers of toxic materials by certain manufacturing facilities available to the public. EPA first published the release data in 1989, and by 1999, reportable emissions had dropped by 46 percent, despite the rapidly growing economy. This trend led EPA to call TRI "one of the most effective environmental programs ever legislated by Congress and administered by EPA."

EPA's success with TRI made information disclosure a core mission throughout the 1990s, and the agency and Congress initiated a number of new disclosure programs. All of these programs were based on the premise that information disclosure works by drawing environmental groups, the press, local communities, peer firms, investors, a range of other actors, and facilities themselves into a complex web of communication and action that pressures low-performing firms to improve environmental performance. These dynamics have been greatly enhanced by advances in information technology and computer networks.

Opponents of the programs, however, charged that they were excessively costly, created a great deal of uncertainty regarding decisions about pollution control, and were subject to abuse by those with ill intent. Tough political battles shaped each disclosure program, but new and expanded programs did emerge.

Following the attacks of September 11, 2001, EPA and other federal departments and agencies decided to remove some disclosed information from their websites and engaged in an extensive review of all public information. Agencies said they feared that such information might provide targeting information for terrorists. The new terrorist threat added strong fuel to the fire of the opponents of disclosure. In the new climate, progress on disclosure programs has stalled, and such programs' potential benefits and harms have come under much closer scrutiny.

> *The new terrorist threat added strong fuel to the fire of the opponents of disclosure. In the new climate, progress on disclosure programs has stalled, and such programs' potential benefits and harms have come under much closer scrutiny.*

Recommendations

I recommend that your administration consider key efforts to reinvigorate information disclosure at EPA and undertake a very public and transparent effort to address the claims of disclosure's proponents and critics.

Enhancing Existing Programs

The following should be top priorities for advances in existing programs:

- *Normalize the TRI for production.* Although it is useful to have information on total volumes of pollution, gaining a more nuanced understanding of pollution trends requires knowing how pollution relates to production. For example, if pollution declines by 10 percent for a facility that has cut production by 50 percent, we would draw a far different conclusion than if pollution declines by 10 percent for a facility where production has expanded by 50 percent. This effort would build on EPA's experience with production normalization in the Sector Facility Indexing Project and extend it to all TRI facilities.
- *Advance efforts to bring risk-related information into TRI.* EPA should continue the advances the agency has made in this area through the Risk Screening Environmental Indicators model, which incorporates information about toxicity and population into TRI's data about releases.
- *Improve the Enforcement and Compliance History On-line (ECHO) database.* EPA's ECHO database (www.epa.gov/echo) integrates compliance data on 800,000 facil-

ities nationwide and is a very big step in making such information easily accessible to the public. EPA should continue to improve ECHO with additional state compliance data (especially for smaller facilities), improving the ability to compare across facilities, sectors, and states, and linking to information on current activities, such as permit applications and renewals.

■ *Improve the ability to link facility-specific information and provide a one-stop shop for facility information.* EPA and states collect a long list of information about what some estimate to be 2 million facilities under various environmental statutes. However, the problems with common facility identifiers and other difficulties hamper data integration. A top priority should be integrating data and making it easily accessible as facility-specific profiles, building on EPA's work on the Facility Registry System, Envirofacts data warehouse, and ECHO.

Creating a New Program

In addition to enhancing existing programs, EPA should consider using information disclosure as a means for advancing its top environmental goals, whatever they may be. For example, if a top priority is improving drinking-water quality, then an integrated public database on water quality test results should be part of any federal initiative. If a top priority is coastal waters, integrating data on coastal water pollution should be considered.

Any new disclosure effort should take advantage of the lessons learned from previous disclosure programs. Progress indicators should come as close as possible to direct information about risks and be readily interpretable, comparable among facilities and geographic areas, and easily updated and monitored. Disclosed information should complement other regulatory activities and provide information that supports effective public participation in decisionmaking. Approaches to disclosure should recognize the role that nongovernmental organizations (NGOs) and the media play as intermediaries, but not discount the potential for proprietary data collection by firms, under some conditions, to spur action.

Disclosed information should complement other regulatory activities and provide information that supports effective public participation in decisionmaking.

Equally important as the substance of new and enhanced information disclosure efforts is the process by which they are developed, given the increased scrutiny after September 11. This process should involve an explicit cataloging of claims made by disclosure's proponents and critics, and a very transparent process of evaluating the claims.

Three claims by proponents and three claims by opponents are likely to dominate the discussion. The following are likely to be the most prominent claims by disclosure's proponents:

■ Disclosure promotes "right to know," whereby individuals, families, and communities are able to access enough information about risks to make informed choices about protection.

■ Disclosure improves environmental performance by compelling facilities to improve environmental management.

- Information collected and shared for the purposes of disclosure generates new knowledge about environmental problems and solutions.

In contrast, the most prominent claims by disclosure's critics are likely to be:

- The cost of information collection and reporting is excessive, particularly when public disclosure increases the need for data quality.
- Disclosure creates a great deal of uncertainty for firms, because an unpredictable public, rather than government managers, determines what facilities should do.
- Disclosure allows unintended use of information, the most salient since September 11 being the threat of terrorism, but also including the use of information by business competitors.

Cautionary Notes

The development of information disclosure programs—and the cataloging and evaluation of claims—should be approached with humility, because many questions remain unanswered. Although the conventional wisdom holds that disclosure can be an effective means of changing environmental behavior, many questions still remain about how much impact disclosure really has, once other forces, such as changes in economic conditions, production, and other factors, are taken into account. Similarly, success from disclosure of one type of information (such as toxic releases to air) may not transfer into success from disclosure of other types of information (such as drinking-water contamination). The means by which disclosure leads to action also is poorly understood. Does it mainly arise from community pressure, from increased regulatory attention, from proactive efforts on the part of firms, or from some combination of all of these?

A new information disclosure effort should be accompanied by research funding on the effectiveness and potential pitfalls of disclosure and an up-front commitment to periodic program evaluation.

Questions about disclosure's costs, uncertainties, and unintended consequences must be considered. Poorly designed programs can create excessive paperwork burdens that add little to program success. Likewise, inappropriate choices about data presentation can inflame public concern unnecessarily; careless data selection may also create privacy or security problems without adding value for a concerned public. We know enough about some of these issues to make reasonable choices about program design. Other issues require further research.

To help answer all of these questions, a new information disclosure effort should be accompanied by research funding, perhaps through cooperation between EPA and the National Science Foundation, on the effectiveness and potential pitfalls of disclosure and an up-front commitment to periodic program evaluation.

Conclusions

Your administration should take the opportunity to reinvigorate information disclosure as a policy tool through advances in TRI and other existing programs, and possibly through new programs that address top environmental priorities of your administration. How these policies are developed will be very important, given the increased scrutiny on disclosure since September 11. The time is ripe for an open and transparent airing and evaluation of the claims made by disclosure's proponents and critics.

T.C.B.

Better Evaluation of Life-Saving Environmental Regulations

by Maureen L. Cropper

Mr. President, I recommend that you issue a new executive order requiring that cost-effectiveness analyses be conducted for all major health and safety regulations, using a uniform set of protocols. Agencies such as the U.S. Environmental Protection Agency, the Food and Drug Administration, and the Department of Transportation, which issue regulations whose primary benefits take the form of reductions in illness, injury, and premature mortality, would be required to calculate the cost per unit of health benefit (such as cost per quality-adjusted life year saved) using identical methodologies. Such an executive order would facilitate a comparison of the cost-effectiveness of health and safety regulations both within and across agencies, and it would serve as a complement to existing benefit–cost analyses of health and safety regulations. Several reasons exist for requiring cost-effectiveness analysis as an adjunct to benefit–cost analysis: it avoids monetization of health benefits, which can be controversial; it provides an alternative method of aggregating health benefits that emphasizes life years saved, rather than private willingness to pay; and the cost per quality-adjusted life year saved, which is commonly employed in the public health arena, provides a simple, transparent way of comparing the efficiency of various health and safety regulations designed to achieve similar objectives. In the long run, expansion of the analytical framework to include cost-effectiveness analysis should lead to more efficient regulation and to greater public acceptance of regulatory decisions.

Background

Executive Order 12866 currently requires that regulatory impact analyses be conducted for all economically significant regulations— those regulations with costs over $100 million per year. The Office of Management and Budget (OMB) suggests that benefit–cost analysis (BCA) should be a primary tool for regulatory impact analysis, and it is the tool that the U.S. Environmental Protection Agency uses in evaluating environmental regulations. In the case of air quality and drinking-water regulations, whose primary benefits are improvements in human health, a BCA calculates the expected reduction in illness and premature mortality associated with the regulation and monetizes these health benefits. Subtracting regulatory costs from monetized benefits provides an estimate of the net benefits of the regulation.

Within EPA, a BCA can inform regulation in three ways:

- It can be used to determine whether the benefits of a proposed regulation outweigh the costs; for example, whether the Tier II Emissions Standards pass the benefit–cost test;
- It can aid in choosing among alternate levels of an environmental standard; for example, to compare the net benefits of reducing the maximum contaminant level for arsenic in drinking water from 50 parts per billion (ppb) to 20, 10, 5, or 3 ppb.
- It can help in setting priorities within a program office; if, for example, a 5 percent reduction in annual average concentrations of fine particulates yields higher net benefits than a 5 percent reduction in annual average maximum one-hour ozone concentrations, then regulations designed to reduce fine particulates may be given higher priority than regulations aimed at reducing ozone.

In practice, the most prominent use of BCA is to examine whether a given regulation passes the benefit–cost test. Recent air quality regulations, such as the 1990 Clean Air Act Amendments, the Tier II Emissions Standards, and the Heavy Duty Engine/Diesel Fuel Rule, have been estimated to yield extremely large net benefits. Indeed, OMB (2003) estimates that four air quality regulations accounted for more than half of the net benefits of 107 health and safety regulations issued between 1992 and 2002. This suggests a second, interagency, use of BCA: comparing net benefits of regulations across agencies suggests where future regulatory efforts might be directed—namely, toward agencies whose regulations have yielded large net benefits.

How would current practice in evaluating environmental regulations be altered by requiring cost-effectiveness analysis (CEA) as a supplement to BCA? Under CEA, reductions in cases of illness and premature mortality would be aggregated by calculating the number of equivalent life years gained, rather than by using weights that reflect what people would pay to avoid illness and reduce risk of death, as is the case with BCA. To illustrate, if an air quality regulation were estimated to save the lives of 100 75-year-olds, this would be translated into the equivalent number of life years saved. If the regulation were estimated to prevent 100 cases of chronic bronchitis (with a given age of onset), this would be translated into an equivalent number of life years saved by determining the relationship between a year of life with chronic bronchitis and a year of life in good health. If a year with chronic bronchitis were equivalent to two-thirds of a year in good health, avoiding a year of chronic bronchitis would save one-third of a life year. Requiring a "reference case" CEA would mean using a uniform set of such weights to aggregate health effects.

Rationale for a Reference Case CEA

CEA health and safety regulations using a uniform set of protocols should be required, rather than continuing to rely on BCA, for at least three reasons. The first stems from difficulties in empirically implementing BCA: specifically, problems exist in obtaining reliable monetary values of the health benefits in a BCA, especially for reductions in risk of death. These problems are avoided by CEA, which does not entail monetization of health outcomes. Second, CEA presents an alternative approach to calculating benefits. BCA is based on the premise that health improvements should be valued by what people would pay to obtain them—an approach that, if applied consistently, means that the value of saving the life of a rich man counts more than saving the life of a poor man. In contrast, CEA counts a year of life saved the same regardless of the economic status of the individual whose life is extended, although the value may depend on the individual's health status. Third, CEA facilitates making comparisons of the cost of health and safety regulations both within and across agencies, which should promote transparency in decisionmaking.

Problems in Monetizing Health Benefits

The net benefits of environmental regulations, especially regarding air quality and drinking-water quality, are extremely sensitive to the value attached to reductions in premature mortality. For example, in BCAs of air quality regulations, the value of reductions in premature mortality typically constitute 90 percent of the monetized benefits of the regulation. Controversies surrounding the value of a statistical life (VSL) have led to controversies in the use of BCA. The VSL is the value of reductions in the risk of dying that, when added up, will save one life. It is the sum of what people are willing to pay for these risk reductions. For example, a regulation that reduces risk of death by 1 in 10,000 for each of 10,000 people will, on average, save one of these lives. If each person would pay $500 for this risk reduction, the VSL is $5 million.

At least two concerns surround estimates of the VSL. First, although economists have made progress in estimating the value of reductions in mortality risk (for example, by examining compensating wage differentials in the labor market), empirical estimates of the VSL remain very fragile—that is, they are very sensitive to the specification of the models used to estimate them. Indeed, in a 2003 study commissioned by EPA to examine the robustness of compensating wage differentials, Dan Black and his coauthors[1] reported that "collectively, these findings lead us to have severe doubts about the usefulness of existing estimates [of compensating wage differentials] to guide public policy."

A second concern is that estimates of the VSL depend primarily on studies of prime-aged workers—men in their forties—but they are used to value reductions in risk of death for much older persons; typically, half of the lives extended by an air quality regulation are those of persons 75 years of age and older, whose remaining life expectancy is much lower than that of workers in compensating wage studies. According to BCA, the lives of older persons should be valued by what these persons are willing to pay for reductions in their risk of death; however, few reliable

Table 25-1

Endpoint	Pollutant	Avoided incidence (cases/year)	Monetary benefits (millions 1999$)
Premature mortality (adults ages 30 and over)	PM	8,300	$62,580
Chronic bronchitis	PM	5,500	$2,430
Hospital Admissions from Respiratory Causes	Ozone and PM	4,100	$60
Hospital Admissions from Cardiovascular Causes	Ozone and PM	3,000	$50
Emergency Room Visits for Asthma	Ozone and PM	2,400	<$5
Acute bronchitis (children, ages 8–12)	PM	17,600	<$5
Upper respiratory symptoms (asthmatic children, ages 9–11)	PM	193,400	$10
Lower respiratory symptoms (children, ages 7–14)	PM	192,900	<$5
Asthma attacks (asthmatics, all ages)	Ozone and PM	361,400	B_a
Work loss days (adults, ages 18–65) (adjusted to exclude asthma attacks)	Ozone and PM	1,539,400	$160
Minor restricted activity days (adults, ages 18–65)	PM	9,838,500	$530
Other health effects	Ozone, PM, CO, NMHC	$U_1+U_2+U_3+U_4$	$B_1+B_2+B_3+B_4$
Decreased worker productivity	Ozone	—	$140
Recreational visibility (86 Class I Areas)	PM	—	$3,260
Residential visibility	PM	—	B_5
Household soiling damage	PM	—	B_6
Material damage	PM	—	B_7
Nitrogen Deposition to Estuaries	Nitrogen	—	B_8
Agriculture crop damage (6 crops)	Ozone	—	$1,120

Source: *Federal Register*, Vol. 66, No. 12, Thursday, January 18, 2001, p. 5105.

estimates exist of the willingness to pay (WTP) of older persons. Simulations of standard economic models (the life-cycle consumption-savings model) suggest that the WTP for mortality risk reduction eventually should decline with age.

Reductions in mortality risk are not the only health effect that is difficult to value. Table 25-1, which summarizes the benefits of a recent air quality regulation, the Heavy Duty Engine/Diesel Fuel Rule, lists the other health effects that usually are quantified in the BCA of an air quality regulation: chronic bronchitis, asthma attacks, and hospital admissions for respiratory illness and cardiovascular disease.

The striking fact that this table illustrates is that these serious illnesses count for very little relative to avoided premature mortality. This is in part due to difficulties in estimating what people would pay to reduce their risk of contracting chronic bronchitis or suffering a heart attack. As a result, a hospital admission for cardiovascular disease is valued by the reduction in medical costs associated with the episode; the value of avoided pain and discomfort are ignored. This has the unintended effect of diminishing the importance of reducing serious illness relative to reducing premature mortality. In the table, the 5,500 avoided cases of chronic bronchitis, each with an average age of onset of 45, count less than 4 percent of the value of extending the lives of 8,300 persons with an average age of 75. A CEA of the same rule gives a very different picture: the reduction in quality-adjusted life years (QALYs) provided by 5,500 avoided cases of chronic bronchitis is more than 40 percent of the QALYs associated with avoided mortality.[2]

Differences in Ethical Perspectives

This leads to the second advantage of CEA as an adjunct to BCA: it provides a different method of aggregating health benefits, doing so by considering healthy time gained rather than private willingness to pay. CEA weights the value of reductions in risk of death by the expected number of life years gained rather than by the dollar amount that an individual would pay for the risk reduction. This implies that saving the life of a 75-year-old, with a remaining life expectancy of 11 years, will be given one-third the weight of saving the life of a 45-year-old, with a remaining life expectancy of 34 years. (This assumes that future life years saved are not discounted, which would reduce the 3:1 ratio.) Current BCAs assign equal weights to saving the two lives. Which approach is correct? The answer is that both approaches have validity and are consistent with individual preferences, depending on how choices are presented to people. When individuals are placed in the role of social decisionmakers and asked how they would allocate medical research dollars to save the lives of young people or of older people, people of all ages—especially those over 65—value young people relative to older persons even more than in proportion to the ratio of their remaining life expectancies.[3]

Is this social perspective valid? In the medical community, a reason frequently given for using CEA to evaluate health programs is that the costs of life extension in this country, under Medicare and Medicaid, are indeed social costs. They are borne in large part by taxpayers rather than by the individuals receiving treatment. Parallels exist in the case of environmental programs. Extending the life of an 85-year-old by reducing air pollution will, in addition to benefiting the individual, likely impose social costs if U.S. taxpayers are paying for the medical and nursing care this person receives. Why should a public health perspective not be examined as well as a private one?

Transparency in Decisionmaking

The third reason for requiring CEA is to promote greater transparency and efficiency in regulation. Listing the cost per QALY, as well as the number of QALYs, saved by a regulation provides a means of comparing regulations that is easy to understand. This would be a useful adjunct to current regulatory impact analyses within agencies and, if cost-effectiveness ratios were calculated using the same methodology, a convenient way of making comparisons across agencies. The point here is not that you should rely exclusively on CEA to make regulatory decisions, but that it provides useful information to supplement existing BCAs and studies of the distribution of benefits and costs of regulation.

Why not make such comparisons using BCA? Currently, cross-agency comparisons of regulatory effectiveness are difficult both because some agencies have rejected BCA as a method of analysis, and because agencies that do perform BCAs often use different values to monetize health endpoints. For example, the Food and Drug Administration regularly employs CEA rather than BCA for its regulatory impact analyses. The Department of Transportation conducts BCAs but uses a lower VSL ($3 million) than does EPA. You could require that BCAs be performed by all agencies using a common methodology; however, such a requirement would have to deal with the problems that arise in attaching a dollar value to health benefits. One advantage of CEA is that it avoids the problem of monetization.

> *A reference case cost-effectiveness analysis would be a useful adjunct to existing regulatory impact analyses conducted by agencies that issue health and safety regulations, including EPA. It would avoid the controversies surrounding monetization of health benefits, provide a public health perspective on health valuation, and facilitate comparing the cost of regulations that deliver health benefits, both within and across agencies.*

The use of CEA also would help inform decisionmaking within agencies. In evaluating air quality regulations within EPA, CEA would provide useful information on the cost-effectiveness of air quality regulations from a health perspective. Currently, BCAs of air quality regulations differ widely in the extent to which nonhealth benefits are quantified. When a regulation yields large health benefits, less attention may be paid to quantifying nonhealth benefits. It would be useful to know, from a health perspective, whether the Heavy Duty Engine/Diesel Fuel Rule is a "better buy" than the Tier II Emissions Standards. A CEA—computed solely using health benefits—would answer this question. This does not imply that you should ignore the nonhealth benefits of either rule, but that it would be useful to compare the rules on an equal footing. The rules also could be compared taking into account the nonhealth benefits that have been quantified by subtracting the value of these benefits from costs before dividing by the number of QALYs saved.

Conclusions

A 2003 guidance by the OMB[4] suggests that CEAs should be conducted for health and safety regulations whenever possible. OMB's guidance, however, stops short of requiring a CEA, or of requiring that a set of uniform procedures be used by all agencies in conducting a CEA. I believe that a reference case CEA would be a useful adjunct to existing regulatory impact analyses conducted by agencies that issue health and safety regulations, including EPA. It would avoid the controversies surrounding monetization of health benefits, provide a public health perspective on health valuation, and facilitate comparing the cost of regulations that deliver health benefits, both within and across agencies.

How, exactly, should your administration use the results of a CEA in conjunction with BCAs? Currently, estimates of the net benefits of regulations provide a useful input into the regulatory process, but they are not used in a rote fashion; regulations may be issued even if their net benefits are small (or even negative), if compelling equity or other considerations exist. The same is true of CEA. No one would suggest that a regulation with a cost per quality-adjusted life year saved in excess of some bright line be rejected solely on the grounds of having a high cost-effectiveness ratio. Efficiency in regulation is, however, an important goal, and providing alternate efficiency measures should ultimately improve the policy process.

M.L.C.

Notes

The findings, interpretations, and conclusions expressed in this paper are entirely those of the author. They do not necessarily represent the views of the World Bank, its executive directors, or the countries they represent.

1. Black, Dan A., Jose Galdo, and Liquin Liu. 2003. *How Robust Are Hedonic Wage Estimates of the Price of Risk?* Unpublished report to U.S. EPA. R 82943001. Syracuse, NY: Syracuse University.

2. Hubbell, Bryan J. 2002. Implementing QALYs in the Analysis of Air Pollution Regulations. Photocopy. Innovative Strategies and Economics Group, U.S. EPA.

3. Cropper, Maureen L., Sema K. Aydede, and Paul R. Portney. 1994. Preferences for Life-Saving Programs: How the Public Discounts Time and Age. *Journal of Risk and Uncertainty* 8: 243–65.

4. OMB (Office of Management and Budget). 2003. *Informing Regulatory Decisions: 2003 Report to Congress on the Costs and Benefits of Federal Regulations and Unfunded Mandates on State, Local and Tribal Entities.* Circular A-4, September 17, 2003. Washington, DC: Office of Management and Budget.

Index

wind power, 21, 28, 30. *See also* renewable
  energy
Wisconsin, 30, 32, 85
working poor, 22, 65–66, 96–97, 112
World Health Organization (WHO), 113

Wyoming, 32, 58

zero-based budgeting, 78, 81
zero-emissions coal plant, 13
zonal management system, 114–119